Number Theory and Its Applications

Edited by Cheon Seoung Ryoo

Published in London, United Kingdom

IntechOpen

Supporting open minds since 2005

Number Theory and Its Applications
http://dx.doi.org/10.5772/intechopen.78090
Edited by Cheon Seoung Ryoo

Contributors
Pedro J. Miana, Natalia Romero, Triantafyllos Konstantinos Makarios, Sudev Naduvath, Samin Riasat, Tomáš Visnyai, Vladimír Baláž, Jung Yoog Kang, Rajesh Cherian Roy, Anant Godbole, Abdelhakim Chillali, Lhoussain El Fadil, Cheon Seoung Ryoo, Aleksandr Shcherbakov Gennadiyevich

First published in London, United Kingdom, 2020 by IntechOpen
IntechOpen is the global imprint of INTECHOPEN LIMITED, registered in England and Wales, registration number: 11086078, 5 Princes Gate Court, London, SW7 2QJ, United Kingdom
Printed in Croatia

British Library Cataloguing-in-Publication Data
A catalogue record for this book is available from the British Library

Additional hard and PDF copies can be obtained from orders@intechopen.com

Number Theory and Its Applications
Edited by Cheon Seoung Ryoo
p. cm.
Print ISBN 978-1-83968-050-2
Online ISBN 978-1-83968-051-9
eBook (PDF) ISBN 978-1-83968-052-6

For EU product safety concerns:
IN TECH d.o.o., Prolaz Marije Krucifikse Kozulić 3, 51000 Rijeka, Croatia, info@intechopen.com or visit our website at intechopen.com.

Meet the editor

Dr. Cheon Seoung Ryoo is a professor in Mathematics at the Hannam University. He received his PhD degree in Mathematics from the Kyushu University. Dr. Ryoo is the author of several research articles in numerical computations with guaranteed accuracy. Also, he has contributed in the field of scientific computing, p-adic functional analysis, and analytic number theory. More recently, he has been working with quantum calculus, special functions, differential equations, and dynamical systems.

Contents

Preface

Number theory and its applications are well known for their proven properties and excellent applicability in interdisciplinary fields of science.

This book contains a collection of selected and refereed papers on the most recent developments in number theory and its applications.

There are so many topics related to number theory that it is hard to list them all, but potential topics include, but are not limited to, the following:

- Algebraic number theory
- Analytic number theory
- Geometric number theory
- Computational number theory
- Transcendental number theory
- Arithmetic combinatorics
- Arithmetic geometry
- Arithmetic topology
- Arithmetic dynamics

The editor of this book sought to cover various aspects of the number theory and its applications in the current project by introducing a variety of welcoming themes from the international audience of mathematicians and researchers. The editor has organized chapters as a result of research by a group of leading researchers in the number theory and related fields. The goal of this book is to provide an overview of the current research in the field of number theory and its applications.

The main subject areas are divided into number theory and its applications.

These include “A new integer-to-integer transform”, “Digit sums and infinite products”, “The Borel-Cantelli lemmas, and their relationship to limit superior and limit inferior of sets”, “Prime numbers distribution line”, “Moments of catalan triangle numbers”, “Modular sumset labelling of graphs”, “Determination of the properties of (p, q)-sigmoid polynomials and the structure of their roots”, “$\mathcal{I}$–convergence of arithmetical functions”, “Identification of eigen-frequencies and mode-shapes of beams with continuous distribution of mass and elasticity and for various conditions at supports”, “Some identities involving 2-variable modified degenerate Hermite polynomials arising from differential equations and distribution of their zeros”, and “Elliptic curve over a local finite ring R_n”.

We hope that this book will be timely and fill a gap in the literature on the number theory and related fields. We also hope that it will promote further research and development in this important field.

Thanks to the authors for their creative contributions and the reviewers for their prompt and careful reviews.

Cheon Seoung Ryoo
Professor,
Hannam University,
South Korea

Section 1

Number Theory

Chapter 1

A New Integer-to-Integer Transform

Rajesh Cherian Roy

Abstract

This chapter presents a detailed analysis of an integer-to-integer transform that is closely related to the discrete Fourier transform, but that offers insights into signal structure that the DFT does not. The transform is analyzed for its underlying properties using concepts from number theory. Theorems are given along with proofs to help establish the salient features of the transform. Two kinds of redundancy exist in the transform. It is shown how redundancy implicit in the transform can be eliminated to obtain a simple form. Closed-form formulas for the forward and inverse transforms are presented.

Keywords: transforms, discrete transforms, integer transforms, DFT, MRT

1. Introduction

Transforms are tools used in signal processing to arrive at deeper insights into the underlying structure of signals. The discrete Fourier transform (DFT), discrete cosine transform (DCT) [1], discrete sine transform (DST), discrete Hartley transform (DHT) [2], and discrete wavelet transform (DWT) are significant discrete transforms. Discrete transforms are characterized by their basis matrices. The Haar transform is distinct in that its basis matrix has only 1, −1, or 0 as elements. The Walsh-Hadamard transform basis matrix is entirely composed of 1 and −1. The discrete Fourier preprocessing transform [3] also has only 1, −1, or 0 in its basis matrix. A new discrete transform, M-dimensional real transform (MRT) based on linear congruences was first proposed [4] for two-dimensional signals. Its basis matrix contains 1, −1, and 0 only. However, it has a high level of redundancy. The orthogonal discrete periodic radon transform (ODPRT) [5] is another transform that is based on linear congruences. MRT has been applied to image compression by making use of a set of unique MRT coefficients [6, 7]. A one-dimensional form of the MRT exists [8]. The Haar transform is related to the Walsh-Hadamard transform [9]. The MRT is related [10] to the Haar transform. One can be obtained from the other through a sequence of bit reversal operations on the rows and columns. Similarly, the Hadamard transform and the MRT can be derived from each other [11]. In [12], 2-D gcd-delta functions that contain only zeroes and ones are used to generate integer 2D DFT pairs. The Weyl transform which has binary valued inner elements is described in [13]. Integer-to-integer approximations of DST are still an active research area, as can be seen in [14]. In comparison to most integer-to-integer transforms, the MRT is distinguished by the utmost simplicity of its kernel. In this context, the MRT can be grouped together with Haar and Walsh-Hadamard transforms as regards the contents of their basis matrices. This chapter presents a detailed study of the MRT in its one-dimensional form and offers an analytical path that justifies placing the MRT beside the Haar transform in the family of discrete transforms.

The MRT is defined and its salient characteristics are presented by way of theorems and their proofs. The redundancy inherent in the MRT is studied and the redundancy-removed version of the MRT named the unique MRT (UMRT) is presented. The inverse UMRT is presented. Described next is a form of the UMRT basis matrix which can be indexed using only 2 indices.

2. Mapped real transform (MRT)

2.1 Forward 1-D MRT

The MRT $Y_k^{(p)}$ of a 1-D sequence x_n, $0 \le n \le N-1$ is defined [7] as

$$Y_k^{(p)} = \sum_{\forall n \Rightarrow ((nk))_N = p} x_n - \sum_{\forall n \Rightarrow ((nk))_N = p+M} x_n$$
$$k = 0, 1, 2 \ldots N-1, \ \ p = 0, 1, 2 \ldots M-1, \text{and} M = \frac{N}{2} \tag{1}$$

In Eq. (1), k can be considered analogous to the frequency in DFT, and p signifies the phase. Thus, 1-D MRT produces M arrays each of size N, given a signal of size N. Hence, MN coefficients have to be computed using only real additions. Another expression for the 1-D MRT [7] is

$$Y_k^{(p)} = \sum_{n=0}^{N-1} A_{k,p,n} x_n, 0 \le k \le N-1, 0 \le p \le M-1 \tag{2}$$

$$A_{k,p,n} = \begin{matrix} 1 \ if [nk]_N = p \\ -1 if \|nk\|_N = p + M \\ 0 \, otherwise \end{matrix} \tag{3}$$

Thus, the kernel $A_{k,p,n}$ maps the data x_n into the 1-D MRT $Y_k^{(p)}$. For example, Let $x = [95 \ 23 \ 61 \ 49 \ 89 \ 76 \ 46 \ 2]$, $N = 8$.
Then, $Y_k^{(p)}$, the corresponding MRT of x, is

$$\begin{aligned} Y_k^{(0)} &= [441 \ \ 6 \ \ 77 \ \ 6 \ \ 141 \ \ 6 \ \ 77 \ \ 6] \\ Y_k^{(1)} &= [0 \ \ -53 \ \ 0 \ \ 47 \ \ 0 \ \ 53 \ \ 0 \ \ -47] \\ Y_k^{(2)} &= [0 \ \ 15 \ \ 48 \ \ -15 \ \ 0 \ \ 15 \ \ -48 \ \ -15] \\ Y_k^{(3)} &= [0 \ \ 47 \ \ 0 \ \ -53 \ \ 0 \ \ -47 \ \ 0 \ \ 53] \end{aligned}$$

2.2 Composition of MRT coefficients

One set of elements that combine to form an MRT coefficient contains those elements satisfying the congruence $(nk)_N = p$, and another set contains those satisfying the congruence $((nk))_N = p + M$. Thus, the two relevant congruences are:

$$((nk))_N = p \tag{4}$$

$$((nk))_N = p + M \quad (5)$$

The group of data elements whose indices satisfy the congruence relation $((nk))_N = p$ is defined as the positive data group or positive set of the 1-D MRT coefficient $Y_k^{(p)}$ The group of data elements whose indices satisfy the congruence relation $((nk))_N = p + M$ is defined as the negative data group or negative set of the 1-D MRT coefficient $Y_k^{(p)}$. For example, the data elements $x_0,\ x_2,\ x_4,\ x_6$ form the positive set of the MRT coefficient $Y_4^{(0)}$ and data elements $x_1,\ x_3,\ x_5,\ x_7$ form the negative set of the MRT coefficient $Y_4^{(0)}$.

An MRT coefficient has two indices, frequency and phase. By formal definition of the MRT, phase has values in the range $[0, M-1]$. Although by definition, $p \in [0, M-1]$, number theory allows $p \in [0, N-1]$. p is referred to as a valid phase for a given value of k if $k|p$. For example, for $N = 6$, if $k = 2$, then $p = 0, 2, 4$ satisfy $k|p$, and hence, these are valid phases for this value of k. A phase p is defined to be an allowable phase index if $p < M$.

2.3 Theorem 1: periodicity

$$Y_k^{(p)} = -Y_k^{(p+M)}$$

Proof. The elements n_a that are in the positive group of the MRT coefficient $Y_k^{(p)}$ can be found as the solutions of

$$((n_a k))_N = p \quad (6)$$

The elements n_a' that are in the negative group of the MRT coefficient $Y_k^{(p)}$ can be found as the solutions of

$$((n_a' k))_N = p + M \quad (7)$$

The elements n_b that are in the positive group of the MRT coefficient $Y_k^{(p+M)}$ can be found as the solutions of

$$((n_b k))_N = p + M \quad (8)$$

The elements n_b' that are in the negative group of the MRT coefficient $Y_k^{(p+M)}$ be found as the solutions of

$$\begin{aligned} &((n_b' k))_N = p + M + M = p + N, \text{which } \textit{can be} \text{ written } \textit{as} \\ &((n_b' k))_N = p \end{aligned} \quad (9)$$

From Eqs. (6) and (9), it can be inferred that

$$n_a = n_b' \quad (10)$$

From Eqs. (7) and (8), it can be inferred that

$$n_a' = n_b \quad (11)$$

From Eqs. (10) and (11) and the definition of MRT in Eq. (1),

$$
\begin{aligned}
&Y_k^{(p)} = \sum n_a - \sum n_b \\
&\sum n'_b - \sum n'_a \\
&-\left(\sum n'_a - \sum n'_b\right) \\
&-Y_k^{(p+M)} \\
&\therefore Y_k^{(p)} = -Y_k^{(p+M)}
\end{aligned}
$$

2.4 Theorem 2: existence conditions

An MRT coefficient $Y_k^{(p)}$ exists for data of order N if either of the following two conditions is satisfied:

Condition 1: $g(k,N)|p$

Condition 2: $g(k,N)|p+M$

If Condition 1 is satisfied, the positive data set is not a null set. If Condition 2 is satisfied, the negative data set of the MRT coefficient is not a null set.

Proof. An MRT coefficient $Y_k^{(p)}$ exists given N if congruences (4) and/or (5) have solutions. The necessary and sufficient condition for (4) to have solutions is that $g(k,N)|p$. Similarly, $g(k,N)|p+M$ becomes the necessary and sufficient condition for (5) to have solutions.

A linear congruence, $((nk))_N = p$, if solvable, has $g(k,N)$ solutions mod N. Hence, there exist $g(k,N)$ solutions for n in the range $[0, N-1]$, and thus, there exist $g(k,N)$ elements in the positive set. Also, if there is a member n_0 of the positive set (particular solution), the other solutions are $n = n_0 + \frac{N}{g(k,N)}t, 0 \leq t < g(k,N)$. These are the other members of the positive set.

The indices of the elements in the positive (or negative) set of an MRT coefficient form an arithmetic progression,

$$n_0 + jg_k, j = [0, g(k,N)-1] \tag{12}$$

where n_0 is the smallest member of the positive (or negative) set, and $g_k = \frac{N}{g(k,N)}$.

2.5 Dependence of phase index on frequency

The existence of 1-D MRT coefficients can be studied for their dependence [15] on the frequency index k.

If k and N are relatively prime, then $g(k,N) = 1$. When $g(k,N) = 1$, the MRT coefficients corresponding to k are in existence for all values of $p \in [0, M-1]$. When k divides N, $g(k,N) = k$. The condition according to Eq. (4) now becomes $k|p$. There are $\frac{N}{k}$ such values of p in $[0, N-1]$. These are $p = 0, k, 2k, ..., N-k$. The condition corresponding to Eq. (5) is now $k|p+M$. When $k|p$, the condition $k|p+M$ has solutions only if k divides M. Thus, an MRT coefficient has both positive and negative sets only if k divides M. Otherwise, only one among the positive or negative sets exists, for a given value of p.

When k divides M, the valid phases for which MRT coefficients have positive sets are $p = 0, k, 2k, ..., N-k$. There are $\frac{N}{k}$ such phases. When $k|M$ is satisfied, negative sets also exist for all these MRT coefficients. There are thus $\frac{M}{k}$ allowable phases and hence $\frac{M}{k}$ MRT coefficients, the phases are $p = 0, k, 2k, \ldots, M - k$. These MRT coefficients have both positive and negative sets simultaneously.

When k is not a divisor of M, for a valid phase, only one among the positive or negative sets exists. The valid phases are an arithmetic series $p = 0, k, 2k, \ldots, N - k$. In this case, there is no integer c that satisfies $p + ck = p + M$, and hence, $Y_k^{(p)} = -Y_k^{(p+ck)}$ cannot be true. Thus, for $p > M$, where p is a valid phase, there is no valid phase index $p - M$. However, the relation $Y_k^{(p)} = -Y_k^{(p-M)}$ is still satisfied. Thus, for $p > M$, there is an allowable but nonvalid phase $p - M$. We can use the relation $Y_k^{(p)} = -Y_k^{(p-M)}$ to express the MRT corresponding to a valid phase index $p > M$ in terms of an allowable phase index. MRT coefficients formed by valid phases have only positive sets. Thus, the positive set of the MRT coefficient with $p > M$ is the negative set of the MRT coefficient with allowable nonvalid phase $p - M$. Hence, a subset of the M allowable phase indices are valid phases, while the other subset is made up of allowable nonvalid phases of the form $p - M$ that are obtained from valid phases $p > M$.

The gap between two successive valid phases is k. Let there be an allowable and numerically smallest valid phase $p_1 < M$, and let the nearest allowable nonvalid phase be p_2. The valid phase corresponding to p_2 is $p_2' = p_2 + M$. Let q be the smallest integer such that $qk > M$. Then,

$$p_2' = p_1 + qk$$

p_2 is the closest allowable nonvalid phase to p_1. p_1 is the lowest allowable valid phase. There is no valid phase $p_1' = p_1 + M$. The next valid phase is hence given by $p_2' = p_1 + qk$, since q is the smallest integer such that $qk > M$

$$\therefore p_2 - p_1 = qk - M$$

Since $k|N$, there is an integer t satisfying $N = tk$. Hence, $\frac{M}{k} = \frac{t}{2}$. Since k does not divide M, $\frac{t}{2}$ is not an integer. For this to be true, t has to be odd. We know $k = \frac{N}{t}$. Since the division of any even number by an odd number produces an even number, k has to be even. Since $k|N$ and k is even, it follows that $\frac{k}{2}|M$.

Hence,

$$M = \frac{dk}{2}, d \text{ being an integer}$$

$$\begin{aligned} p_2 - p_1 &= qk - \frac{dk}{2} \\ &= \frac{2qk}{2} - \frac{dk}{2} \\ &= (2q - d)\frac{k}{2} \end{aligned}$$

The value of $(2q - d)$ cannot be larger than 1, since, in that case,

$$p_2 - p_1 \text{ is greater than or equal to } k.$$

The distance between p_2' and M has to be lesser than k. Thus, the distance between p_1 and p_2 has to be lesser than k. Hence,

$$p_2 - p_1 = \frac{k}{2} \tag{13}$$

The next valid phase after p_1 is

$$p_3 = p_1 + k \tag{14}$$

From Eqs. (13) and (14),

$$p_3 - p_2 = \frac{k}{2}$$

Let p'_4 be the next valid phase index after p'_2, then

$$p'_4 = p'_2 + k \tag{15}$$

p'_4 has a corresponding nonvalid allowable phase p_4 given by

$$p_4 = p'_4 - M \tag{16}$$

From Eqs. (15) and (16),

$$p_4 = p_2 + k$$
$$\therefore p_4 - p_3 = \frac{k}{2}$$

Hence, there exists an allowable nonvalid phase between every consecutive pair of allowable valid phases. The MRT coefficient produced by these allowable nonvalid phases will have equal magnitude and opposite sign as the MRT coefficient produced by the corresponding non-allowable valid phases.

The sequence of allowable phase indices would thus be:

$$p_0, p_0 + \frac{k}{2,} p_0 + k, p_0 + 3\frac{k}{2} \ldots\ldots\ldots\ldots\ldots\ldots, p_0 + M - \frac{k}{2} \tag{17}$$

When k does not divide M, MRT coefficients exist for these allowable phase indices and they will have either a positive group or a negative group only. There are $\frac{N}{k}$ such allowable phases, and when k does not divide M, $\frac{N}{k}$ is odd. Since $k|N$, the condition for existence for solutions is $k|p$. The smallest value of p that satisfies this condition is $p = 0$. Hence, the first valid allowable phase is $p_0 = 0$. MRT coefficients having only a positive set have allowable phases that are even multiples of $\frac{k}{2}$, starting from 0 and ending at $\frac{N-k}{2}$. There are $\frac{N}{2k} + \frac{1}{2}$ such MRT coefficients. Similarly, the MRT coefficients having only a negative set have allowable phases that are odd multiples of $\frac{k}{2}$, starting from $\frac{k}{2}$ and ending at $M - k$. There are $\frac{N}{2k} - \frac{1}{2}$ such MRT coefficients. The total number of 1-D MRT coefficients is thus $\frac{N}{k}$, and the sequence of allowable phases is:

$$0, \frac{k}{2,} k, \ 3\frac{k}{2,} \ldots\ldots\ldots\ldots\ldots\ldots, M - \frac{k}{2} \tag{18}$$

The sequence of allowable phases that correspond to MRT coefficients having only positive sets is

$$0, k, 2k, \ldots\ldots\ldots\ldots\ldots, M - \frac{k}{2} \tag{19}$$

and the sequence of allowable phases that correspond to MRT coefficients having only negative sets is

$$\frac{k}{2,} 3\frac{k}{2,} 5\frac{k}{2,.} \ldots\ldots\ldots\ldots\ldots, M - k \tag{20}$$

2.6 Theorem 3: index of first element

a. When $k|N$, the index n_a of the first element in the positive data group of an MRT coefficient $Y_k^{(p)}$ is $n_a = \frac{p}{k}$.

b. For $k|M$, if an element with index n belongs in the positive data set of an MRT coefficient $Y_k^{(p)}$, then the element with index $n + \frac{M}{k}$ occurs in the negative data set of the same MRT coefficient.

c. The first element in the positive set of an MRT coefficient $Y_k^{(p)}$ when k does not divide M, has the index $n_a = \frac{p}{k}$. The first element in the negative set of an MRT coefficient $Y_k^{(p)}$, when k does not divide M, is $n_b = \frac{p+M}{k}$.

Proof.

a. The first element in MRT coefficient $Y_k^{(p)}$ will have an index that is the smallest solution of (4),

$$((nk))_N = p.$$

Solutions exist for Eq. (4) only if $g(k, N)$ divides p. Since $g(k, N) = k$, the former condition can be written as $k|p$.

Since $k|p$, the smallest solution to Eq. (4) is $n_a = \frac{p}{k}$, and thus $n_a = \frac{p}{k}$ is the first member in the positive data set of MRT coefficient $Y_k^{(p)}$.

The condition $g(k, M) = k$ is necessary for both positive and negative sets to be present in an MRT coefficient.

b. Given $((nk))_N = p$,

$$\left[\!\left[\left(n + \frac{M}{k}\right)k\right]\!\right]_N = [\![nk]\!]_N + \left[\!\left[\frac{M}{k}k\right]\!\right]_N = p + M$$

Hence, if index n belongs in the positive set, index $n + \frac{M}{k}$ will be present in the negative set since $\left[(n + \frac{M}{k})k\right]_N = p + M$.

c. From (18), when k does not divide M, the sequence of allowable phase indices is

$$0, \frac{k}{2,}k, \ 3\frac{k}{2,} \ldots\ldots\ldots\ldots\ldots, M - \frac{k}{2}$$

From (19), the sequence of allowable phase indices that correspond to MRT coefficients having only positive groups is

$$0, k, 2k, \ldots\ldots\ldots\ldots\ldots\ldots, M - \frac{k}{2}$$

From (20), the sequence of allowable phase indices that correspond to MRT coefficients having only negative groups is

$$\frac{k}{2,} 3\frac{k}{2,} 5\frac{k}{2,.} \ldots\ldots\ldots\ldots\ldots, M - k$$

The first data element that satisfies $(nk)_N = p$ is $\frac{p}{k}$ since $k|p$ is satisfied, p being a valid phase index. Hence, the first element in the positive set has index $n_a = \frac{p}{k}$.

Also, the allowable phase indices that produce MRT coefficients having only negative sets are actually nonvalid allowable phase indices. If p_b is a nonvalid phase index, there is a valid phase index p_a given by $p_a = p_b + M$. Hence, the first data element in the negative set has the index $n_b = \frac{p+M}{k}$.

If k does not divide N, then $n_a = \frac{p}{g(k,N)}$ cannot be a solution since $g(k,N)|k$. The extended Euclidean algorithm can be used to find a particular solution for this case.

2.7 Closed-form expression for 1-D MRT

By the MRT definition, and looking at element indices of both positive and negative data sets in (12), and, assuming n_a belongs in the positive set, and n_b in the negative set, a 1-D MRT coefficient can be expressed in the following manner:

$$\begin{aligned} Y_k^{(p)} = & \left[x_{n_a} + x_{n_a+g_k} + x_{n_a+2g_k} + x_{n_a+3g_k} + \ldots\ldots\ldots + x_{n_a+g(k,N)-1g_k}\right] \\ & - \left[x_{n_b} + x_{n_b+g_k} + x_{n_b+2g_k} + x_{n_b+3g_k} + \ldots\ldots\ldots + x_{n_b+g(k,N)-1g_k}\right] \end{aligned}$$

This can be written as

$$Y_{k_{(p)}} = \sum_{j=0}^{g(k,N)-1} x_{n_a+jg_k} - x_{n_b+jg_k} \tag{21}$$

Here, Eq. (21) is a closed-form formula for MRT coefficients.

When k divides N, using Theorem 3(a), we observe that the first element in the positive set is $n_a = \frac{p}{k}$. Also, using Theorem 3(b), any index in the positive set is related to any index in the negative set, given $k|M$. This relation is $n_b = n_a + \frac{M}{k}$ When k does not divide M, from Theorem 3(c), we obtain the phases that correspond to MRT coefficients with only positive sets, and MRT coefficients with only negative sets.

When $k|M$, $n_a = \frac{p}{k}$, and $n_b = n_a + \frac{M}{k}$. Also, k does not divide M. Using these, we can rewrite Eq. (21) as,

$$\begin{aligned} Y_k^{(p)} = & \left[x_{\frac{p}{k}} + x_{\frac{p}{k}+\frac{N}{k}} + x_{\frac{p}{k}+\frac{2N}{k}} + x_{\frac{p}{k}+\frac{3N}{k}} + \ldots\ldots\ldots + x_{\frac{p}{k}+\frac{(k-1)N}{k}}\right] \\ & - \left[x_{\frac{p}{k}+\frac{N}{2k}} + x_{\frac{p}{k}+\frac{3N}{2k}} + x_{\frac{p}{k}+\frac{5N}{2k}} + x_{\frac{p}{k}+\frac{7N}{2k}} + \ldots\ldots\ldots + x_{\frac{p}{k}+\frac{(2k-1)N}{2k}}\right] \end{aligned} \tag{22}$$

On further simplification,

$$Y_k^{(p)} = \sum_{j=0}^{k-1} \left[x_{\frac{jN+p}{k}} - x_{\frac{(2j+1)N+2p}{2k}}\right]$$

$$p = 0, k, 2k, \ldots, M - k \tag{23}$$

When k does not divide M, it has been seen that positive and negative groups cannot exist together for the same MRT coefficient. For certain values of p, only positive groups exist. For other values of p, only negative groups exist. As seen, for positive groups, $n_a = \frac{p}{k}$, and for negative groups, $n_b = \frac{p+M}{k}$. An MRT coefficient with only a positive group has the following form:

$$Y_k^{(p)} = \left[x_{\frac{p}{k}} + x_{\frac{p}{k}+\frac{N}{k}} + x_{\frac{p}{k}+\frac{2N}{k}} + x_{\frac{p}{k}+\frac{3N}{k}} + \ldots\ldots + x_{\frac{p}{k}+\frac{(k-1)N}{k}}\right]$$

which, when simplified, becomes

$$Y_k^{(p)} = \sum_{j=0}^{k-1} \left[x_{\frac{jN+p}{k}}\right]$$
$$p = 0,\ k,\ 2k, \ldots, M - \frac{k}{2} \tag{24}$$

Similarly, an MRT coefficient with only a negative group has the following form:

$$Y_k^{(p)} = -\left[x_{\frac{p+M}{k}} + x_{\frac{p+M}{k}+\frac{N}{k}} + x_{\frac{p+M}{k}+\frac{2N}{k}} + \ldots\ldots + x_{\frac{p}{k}+\frac{(k-1)N}{k}}\right]$$

which, when simplified, becomes

$$Y_k^{(p)} = -\sum_{j=0}^{k-1} \left[x_{\frac{jN+p+M}{k}}\right]$$
$$p = \frac{k}{2}, \frac{3k}{2}, \ldots, M - k \tag{25}$$

When k and N are co-prime, $g(k, N) = 1$. Hence, the positive set has only one element and similarly the negative set too has only one element. The values of n_a and n_b have to be computed using the Euclidean algorithm or by the trial-and-error method. The MRT coefficient has the following form

$$Y_k^{(p)} = x_{n_a} - x_{n_b}$$
$$p = 0, 1, 2, 3, \ldots\ M - 1 \tag{26}$$

2.8 Physical significance

An MRT coefficient possesses both frequency and phase. A DFT coefficient has only the frequency index. Hence, the presence of an extra index distinguishes the MRT coefficient. The physical significance of an MRT coefficient is that the phase specifies the beginning of the frequency cycle. The MRT can thus be thought of as a time-frequency representation of a 1-D signal. In contrast to the DFT, the MRT while related to the DFT, has localization in both time and frequency. Also, MRT coefficients can be considered to be constituent parts of the DFT; these parts, if weighted by the exponential kernel, would produce the DFT. For $N = 8$, the DFT coefficient Y_k can be expressed in terms of associated MRT as

$$Y_k = Y_k^{(0)} W_8^0 + Y_k^{(1)} W_8^1 + Y_k^{(2)} W_8^2 + Y_k^{(2)} W_8^3$$

3. Redundancy in MRT

For some values of N, a set of MRT coefficients having different values for frequency and phase have the same magnitude. The polarity of the coefficients may be different. This phenomenon is an indication of redundancy in MRT. The MRT can be relieved of the redundancy to arrive at a simpler transform that has no redundancy.

3.1 Theorem 4: complete redundancy

Given $Y_k^{(p)}$, for all h such that $g(h, N) = 1$

$$Y_{(hk)_N}^{((hp)_N)} = Y_k^{(p)} \text{ for } [hp]_N < M, \text{and,}$$
$$Y_{(hk)_N}^{((hp)_N - M)} = -Y_k^{(p)} \text{for } [hp]_N \geq M$$

Proof. From a basic theorem [16] in number theory, if

$$[\![q]\!]_N = d \tag{27}$$

and h is a multiplication factor, then

$$[\![hq]\!]_{\frac{N}{g(h,N)}} = hd \tag{28}$$

If $g(h, N) = 1$, Eq. (28) becomes

$$[\![hq]\!]_N = hd \tag{29}$$

Given $Y_k^{(p)}$, and a set of indices n that satisfies

$$[\![nk]\!]_N = p \tag{30}$$

and a set of indices n' that satisfies

$$[\![n'k]\!]_N = p + M \tag{31}$$

If there is h such that $g(h, N) = 1$, using Eqs. (27)–(29),

$$[\![n(hk)]\!]_N = hp \tag{32}$$

and

$$[\![n'(hk)]\!]_N = h(p + M) \tag{33}$$
$$[\![n'(hk)]\!]_N = hp + hM \tag{34}$$

Since $g(h, N) = 1$, h is odd, and hence

$$hp + hM \equiv hp + M \tag{35}$$

Using Eq. (35), Eq. (34) may be written as

$$[\![n'(hk)]\!]_N = hp + M \tag{36}$$

From the definition of MRT, and Eqs. (32) and (36), it is seen that the set of indices n and n' form the MRT coefficient $Y_{(hk)_N}^{(hp)_N}$. Using Eqs. (30)–(32) and (36),

$$Y_{(hk)_N}^{((hp)_N)} = Y_k^{(p)} \tag{37}$$

If $(hp)_N \geq M$, then

$$Y_{(hk)_N}^{((hp)_N - M)} = -Y_k^{(p)} \tag{38}$$

since

$$Y_k^{(p)} = -Y_k^{(p-M)} = -Y_k^{(p+M)}$$

Hence, the theorem on complete redundancy has been proved.

From Theorem 4, we see that prediction of redundancy is possible from the knowledge of N, frequency and phase. Given $(k,\ p)$, frequency and phase pairs of redundant MRT coefficients are given by Theorem 4. The redundancy condition is that the factor of multiplication that connects the frequencies of redundant MRT coefficients should be co-prime to N. As an example, for $N = 6$, MRT coefficients of $k = 5$ are completely redundant MRT coefficients of $k = 1$, since 1 and 5 are co-prime to 6.

If $k' = hk$ and $g(h, N) = 1$, then MRT coefficients of the two frequency indices k and k' are redundant.

When $k = 1$, MRT coefficients having frequency k' that are co-prime to N can be obtained from MRT coefficients with a frequency of $k = 1$. If $k'|N$, then we cannot obtain k' by multiplying an integer k with an integer h, $k' = hk$, satisfying $g(h, N) = 1$ and $k \neq k'$ Thus, MRT coefficients with frequency indices that are divisors of N cannot be redundant to each other.

Let k′ satisfy $k' = hk$ such that $g(h, N) = 1$ and $k \neq k'$. If k' is nonprime, there is a nontrivial common divisor with N. So, even though it is not a divisor of N, it will have complete redundancy with the common divisor.

3.2 Theorem 5: redundancy frequency groups

MRT coefficients with frequency indices that have common gcd w.r.t. N are all completely redundant to each other.

Proof. Assume k_1 and k_2 are two frequency indices that have common gcd w.r.t. N.

$$g(k_1, N) = k \tag{39}$$

$$g(k_2, N) = k \tag{40}$$

Assume h exists such that

$$g(h, N) = 1 \tag{41}$$

From Eq. (41), h and N are co-prime. Hence,

$$g(hk_1, N) = k \tag{42}$$

$$[\![hk_1]\!]_N = hk_1 - Nq \ \text{ if } \ 0 \leq hk_1 - Nq < N, q \in \mathbb{Z} \tag{43}$$

Using Eq. (43) and gcd property,

$$g\big((hk_1)_N, N\big) = g(hk_1 - Nq, N) = g(hk_1, N) = k \tag{44}$$

From Eqs. (40) and (44),

$$k_2 = [\![hk_1]\!]_N \tag{45}$$

From Eqs. (41) and (45), and using Theorem 4, it can be concluded that frequency indices k_1 and k_2 are completely redundant. Hence, the theorem is proved.

For example, if $N = 8$, the following relations exist:

$$\begin{aligned} Y_1^{(0)} &= Y_3^{(0)} = Y_5^{(0)} = Y_7^{(0)} \\ Y_1^{(1)} &= Y_3^{(3)} = -Y_5^{(1)} = -Y_7^{(3)}, \\ Y_1^{(2)} &= -Y_3^{(2)} = Y_5^{(2)} = -Y_7^{(2)} \text{ and,} \\ Y_1^{(3)} &= Y_3^{(1)} = -Y_5^{(3)} = -Y_7^{(1)}. \end{aligned}$$

Since $k = 1, 3,$ 5 and 7 share the same gcd of 1 w.r.t. N, there is redundancy among the MRT coefficients of these frequencies. Similarly, since $g(2, 8) = g(6, 8) = 2$, $Y_2^{(0)} = Y_6^{(0)}$, and $Y_2^{(2)} = -Y_6^{(2)}$.

Theorem 5 shows that we can group frequencies based on their gcd w.r.t. N. AII nondivisor frequency indices are related to divisor frequency indices through multiplication factors h such that $g(h, N) = 1$. The count of the possible factors of multiplication involved in complete redundancy is Euler's totient function $\Phi(N)$, which specifies the count of positive integers smaller than or equal to N that are co-prime to N, 1 being considered co-prime to every other integer. Given 1-D MRT coefficients having frequency $k = 1$, there would be $\Phi(N) - 1$ other frequencies whose MRT coefficients can be obtained from this MRT coefficient. Combined with $k = 1$, these $\Phi(N)$ frequencies thus comprise the group of frequencies that satisfy $g(h, N) = 1$. There are other sets of frequency indices that have a common gcds w.r.t. N.; and each group corresponds to a some divisor of N. There are $\Phi(N)$ possible factors of multiplication that produce other members of the group of frequency indices associated with k. From Theorem 4, the equation for complete redundancy is $k' = (hk)_N$, where $g(h, N) = 1$. Also,

$$[\![hk]\!]_N = \left[\!\!\left[\left(h + \frac{N}{k}\right)k\right]\!\!\right]_N \tag{46}$$

Eq. (46) implies that the set of k' that is generated from divisor k is unique only for multiplicative factors in the set $\left[0, \frac{N}{k} - 1\right]$, and repeats thereafter for the remaining sets of the same length. Hence, the problem now gets reduced to $k' = ((hk))_{N/k}$ where $g\left(h, \frac{N}{k}\right) = 1$. The number of such multiplicative factors h is $\Phi\left(\frac{N}{k}\right)$, and these factors are the totatives of $\frac{N}{k}$ Hence, the number of frequency indices that are related by complete redundancy to a frequency index k is given by $\Phi\left(\frac{N}{k}\right)$ and they are obtained by $k' = (hk)_N$ where $g\left(h, \frac{N}{k}\right) = 1$.

From number theory [17],

$$\sum_{d\,VN} \Phi(d) = N, \quad \text{if } N \geq 1, \tag{47}$$

From Eq. (47), the sum of the terms $\Phi\left(\frac{N}{k}\right)$ over all divisors of N is given by N, since $\sum_{d|N}\Phi(d) = \sum_{d|N}\Phi\left(\frac{N}{d}\right)$. Hence, all the N frequency indices $k = [0, N-1]$ have been mapped.

Thus, all the frequency indices of 1-D MRT can be classified on the basis of their gcd w.r.t. N.

3.3 Theorem 6: Mapping between phase indices

One-to-one mapping exists between phases of 1-D MRT coefficients having frequencies k and k' connected through complete redundancy.

Proof. For two MRT coefficients having frequencies k and k' connected by complete redundancy, the number of phases corresponding to each frequency is equal since $g(k,N) = g(k',N)$, and the number of phase indices is given by $\frac{N}{g(k,N)}$. The phases lie in the range $[0, M-1]$. From Theorem 4 on complete redundancy, the relation between phases is given by $p' = [\![hp]\!]_N$, $g(h,N) = 1$. Using a theorem on the reduced residue systems, on multiplication with h, the resulting group of phase indices p' too have the same elements as the original group. Multiplication of phases in $[0, M-1]$ by h and then computing modulo w.r.t. M produces the same group, but with the order possibly altered. Thus, one-to-one mapping exists between the phases of 1-D MRT coefficients having frequencies k and k' connected through complete redundancy.

To look at an example, when $N = 8$,

$$\begin{aligned} Y_1^{(0)} &= Y_3^{(0)}, \\ Y_1^{(1)} &= Y_3^{(3)} \\ Y_1^{(2)} &= -Y_3^{(2)}, \text{and}, \\ Y_1^{(3)} &= Y_3^{(1)}. \end{aligned}$$

Since $k = 1$ and $k = 3$ are redundant through the co-prime $h = 3$, the set of phase indices of $k = 1$, $p = [0,1,2,3]$, when subjected to the operation $p' = [hp]_N$ would result in $p' = [0,3,6,1]$, which reduces to $p' = [0,3,2,1]$ after the condition $[\![hp]\!]_N < M$ is checked and relevant sign change. Hence, $p = [0,1,2,3]$ maps to $p' = [0,3,2,1]$.

3.4 Derived redundancy

For $N = 6$,

$$Y_3^{(0)} = Y_1^{(0)} - Y_1^{(1)} + Y_1^{(2)}$$

Certain MRT coefficients can be obtained using the summation of certain other unique MRT coefficients. This phenomenon is also a kind of redundancy, but it cannot be considered as complete redundancy. This phenomenon is referred to as derived redundancy. An MRT coefficient is considered a derived MRT coefficient if we can be obtain it by combining other MRT coefficients.

Let $Y_{k'}^{(p)}$ and $Y_k^{(p_a)}$ be the two MRT coefficients. The congruence relations for $Y_k^{(p_a)}$ are $[\![nk]\!]_N = p_a$ and $[\![nk]\!]_N = p_a + M$. The congruence relations for $Y_{k'}^{(p)}$ are $[\![nk']\!]_N = p$ and $[\![nk']\!]_N = p + M$. Assume that a relation $k' = dk$ exists between k and k'.

N = 6	Complete redundancy (5)	Derived redundancy (3)
1	5	3
2	4	6
3		
6		

Table 1.
Complete redundancy and derived redundancy relations for N = 6.

N = 24	Co-primes							Odd divisors
	5	7	11	13	17	19	23	3
1	5	7	11	13	17	19	23	3
2	10	14	22					6
3	15	21	9					
4	20							12
6		18						
8	16							24
12								
24								

Table 2.
Complete redundancy and derived redundancy relations for N = 24.

Hence, $[\![dnk]\!]_N = p$ and $[\![dnk]\!]_N = p + M$. $[\![dnk]\!]_N = p$ may be written as $[\![nk]\!]_N = \frac{p}{d}$ if $g(d, N) = d$, and $d|p$. In other words, d should be a divisor of N. From $[\![nk]\!]_N = \frac{p}{d}$ and $[\![nk]\!]_N = p_a$, $p_a = \frac{p}{d}$. Multiplying both sides of $[\![nk]\!]_N = p_a + M$ *by d*, $[\![dnk]\!]_N = dp_a + dM$. If d is odd, this can be written as $[\![dnk]\!]_N = dp_a + M$, which becomes $[\![nk']\!]_N = p + M$. Hence, provided there exists an odd-valued divisor d of N, such that $k' = dk$, then there is derived redundancy between and $Y_{k'}^{(p)}$ and $Y_k^{(p_a)}$. Hence, we can conclude that there cannot be derived redundancy when N is a power of 2 since N has no divisors that are odd. For even N not a power of 2, derived redundancy exists since N then has divisors that are odd. Details of derived redundancies are shown in **Tables 1** and **2** for $N = 6$ and $N = 24$, respectively.

4. Unique MRT

Based on the two types of redundancy presented above, MRT coefficients can be considered to be unique or relatively unique. It is impossible to obtain MRT coefficients of divisor frequencies from MRT coefficients of other divisor frequencies by complete redundancy. These can hence be named as unique MRT coefficients. If divisors have a relationship to other divisors by multiplication using a divisor that is odd-valued, derived redundancy is exhibited by them. They are thus considered only relatively unique. For $N = 6$, the set of divisor frequencies is {1, 2, 3, 6}. In this set, 3 and 6 are relatively unique divisor frequencies. If we remove these, we obtain the absolutely unique divisor set {1, 2}. It is not possible to express absolutely unique divisors as $k' = dk$, for d an odd integer. Considering the divisor set of N, this condition is satisfied only by divisors that are powers of 2, as observed in the above example. If $k' = 2^a$, then given any k and odd-valued d except $d = 1, \;\; k' \neq dk$. So, the unique divisor set for N contains only divisors which are powers of 2. MRT coefficients formed by unique divisor frequencies are referred to as unique MRT (UMRT) coefficients. Thus, 1-D UMRT is the set of all MRT coefficients having frequencies that are powers of 2.

4.1 Number of unique coefficients

In total, there are MN MRT coefficients given a signal of size N. Now arises the question of the exact number of UMRT coefficients.

If N is a power of 2, the frequencies that form unique coefficients are themselves powers of 2. Since, $k = 0 = [\![N]\!]_N$ and N a power of 2, then $k = 0$ produces a unique coefficient. The first frequency index is $k = 1$, and followed by $k = 2, 4, \ldots, N$. The number of unique coefficients produced by each frequency index k is given by the number of allowable phase indices for each frequency index k. For an MRT coefficient $Y_k^{(p)}$ to exist, p should be divisible by k. When $k = N$, this condition has only one solution for $p, \;\; p = 0$. All other frequency indices have $\frac{M}{k}$ coefficients each. Hence, the total number of UMRT coefficients is given by

$$\begin{aligned}
Tot &= 1 + \sum_{t=0}^{\log_2 M} \frac{M}{2^t} \\
&= 1 + M \sum_{t=0}^{\log_2 M} \frac{1}{2^t} \\
&= 1 + M \frac{1 - \left(\frac{1}{2}\right)^{\left(\log_2 M + 1\right)}}{\frac{1}{2}} \\
&= 1 + N\left(1 - 2^{-\left(\log_2 M + 1\right)}\right) \\
&= 1 + N\left(1 - \frac{1}{N}\right) \\
&= N
\end{aligned}$$

Hence, the number of UMRT coefficients, when N is a power of 2, is N.

If N is not a power of 2, $k = 0$ does not produce UMRT coefficients. Let k' be the highest power of 2 frequency and a divisor of N. Hence $N = dk', \;\; d$ being an odd integer; if d has an even value, then it would violate the assumption that k' is the highest power of 2 frequency and a divisor of N. Since d has an odd value, it is not possible for k' to be a divisor for M as $\frac{d}{2}$ cannot be integer-valued. The number of valid phase indices is $\frac{N}{k'}$ and thus there exist $\frac{N}{k'}$ MRT coefficients in the case that k' does not divide M. AII the numbers that are powers of 2 and divide N and are smaller than k' will be divisors of M. These are $k = 1, 2, 4, \;\; \ldots, \;\; \frac{k'}{2}$. There are $\frac{M}{k}$ MRT coefficients produced by these frequencies. When N is not a power of 2, frequency $k = 0$ can be obtained by derived redundancy. A frequency k =0 is equivalent to $k = N$, since $[\![N]\!]_N = 0$. Frequency N relates to k' through $N = dk', \; d$ being odd.

This can be recalled to be sufficient for derived redundancy to occur. Thus, in case if N is not a power of 2, $k = 0$ does not form an absolutely unique MRT coefficient. Hence, the total number of UMRT coefficients is

$$Tot = 1 + \sum_{t=0}^{\log_2 M} \frac{M}{2^t}$$

$$= 1 + M \frac{1 - \left(\frac{1}{2}\right)^{(\log_2 M + 1)}}{\frac{1}{2}}$$

$$= \frac{N}{k'} + N\left(1 - \frac{2}{2k'}\right)$$

$$= N$$

We thus conclude that N UMRT coefficients are needed to represent a 1-D signal of length N, whether N is a power of 2 or not, and these coefficients are produced by frequencies that are divisors of N and powers of 2, beginning with $k = 1$.

The 1-D UMRT coefficients can be computed as below:

(i) N a power of 2

$$Y_0^{(0)} = \sum_{n=0}^{N-1} x_n \tag{48}$$

$$Y_k^{(p)} = \sum_{j=0}^{k-1} \left(x_{\frac{jN+p}{k}} - x_{\frac{(2j+1)N+2p}{2k}} \right)$$

$$k = 2^t, 0 \le t \le \log_2 M, p = tk, 0 \le t \le \frac{M}{k} - 1 \tag{49}$$

(ii) N not a power of 2

$$Y_k^{(p)} = \sum_{j=0}^{k-1} \left(x_{\frac{jN+p}{k}} - x_{\frac{(2j+1)N+2p}{2k}} \right)$$

$$k = 2^t, 0 \le t \le \log_2 M, p = tk, \quad 0 \le t \le \frac{M}{k} - 1 \tag{50}$$

$$Y_{k'}^{(p)} = \sum_{j=0}^{k'-1} \left(x_{\frac{jN+p}{k}} \right)$$

$$p = tk', \quad 0 \le t \le \frac{M}{k'} - \frac{1}{2} \tag{51}$$

$$Y_{k'}^{(p)} = -\sum_{j=0}^{k'-1} \left(x_{\frac{jN+p+M}{k'}} \right)$$

$$p = tk' + \frac{k'}{2}, \quad 0 \le t \le \frac{M}{k'} - \frac{3}{2} \tag{52}$$

where k' is the highest frequency index that is a power of 2 and also a divisor of N.

5. Inverse MRT

5.1 Theorem 7: N a power of 2

Given the UMRT of a 1-D signal of size N, N being a power of 2, the 1-D signal can be reconstructed from its UMRT by the following formula

$$x_n = \frac{1}{N} Y_0^{(0)} + \sum_{t=0}^{\log_2 M} \frac{1}{2^{t+1}} Y_{2^t}^{\left((2^t n)_N\right)}, \quad 0 \le n \le N-1 \tag{53}$$

Proof. The data element that needs to be recovered from the UMRT is given by x_n. For any frequency index k, the value of the phase index p of the UMRT coefficient $Y_k^{(p)}$ that contains x_n is given by nk $(nk)_N = p$. Thus for a frequency that is a power of 2, $k = 2^a$, the UMRT coefficient that contains x_n is $Y_{2^a}^{\left((2^a n)_N\right)}$. The UMRT coefficient $Y_0^{(0)}$ contains all the elements of the data including x_n since $(nk)_N = 0, \forall n$, when $k = 0$. In Eq. (53), $Y_0^{(0)}$ has a factor of multiplication $\frac{1}{N}$, and the remaining UMRT coefficients have a factor of multiplication $\frac{1}{2^{t+1}}$. Consequently, the x_n that is a part of these coefficients have the corresponding factors of multiplication. The resultant multiplication factor f for x_n due to the summation is obtained as a sum of the individual factors of multiplication.

$$\begin{aligned}
f &= \frac{1}{N} + \sum_{t=0}^{\log_2 M} \frac{1}{2^{t+1}} \\
&= \frac{1}{N} + \sum_{t=0}^{\log_2 M} 2^{-(t+1)} \\
&= \frac{1}{N} + \frac{1}{2}\frac{\left(1 - 2^{-\left(\log_2 M + 1\right)}\right)}{\frac{1}{2}} \\
&= \frac{1}{N} + 1 - \frac{1}{N} \\
&= 1
\end{aligned}$$

Hence, as a result of the summation, one of the components of the result is the data x_n.

A UMRT coefficient $Y_{2^a}^{\left((2^a n)_N\right)}$ contains other terms besides x_n. For the inverse transform formula to be correct, these other terms that occur in the various UMRT coefficients $Y_{2^a}^{\left((2^a n)_N\right)}$ need to get canceled off. It can be proved that they vanish. Let k be the smallest frequency at which any other data elements co-occur with x_n. $Y_0^{(0)}$ can be excluded as all data elements co-occur in it with a positive sign. Leaving out $Y_0^{(0)}$, another element shows co-occurrence with x_n for the first time for $k = 1$. For example, for $N = 8$, $Y_1^{(0)} = x_0 - x_4$. Hence, if x_0 is the element to be obtained, it is seen that x_4 occurs with an opposite sign along with x_0 in the MRT coefficient corresponding to $k = 1$, given $by\, Y_1^{(0)}$. From Eq. (12), when $k = 1$, $g(k, N) = 1$ and so both positive and negative data sets contain only one element each. Using

Theorem 3(b), the distance between corresponding elements in the two data groups is $\frac{M}{k}$. Hence, x_n and x_{n+M}, occur having opposite polarity, in any MRT coefficient of frequency $k = 1$. In the same way, for $k = 2$, the data element x_{n+M}occurs with positive sign since $Y_2^{(p)} = x_n - x_{n+\frac{N}{4}} + x_{n+\frac{N}{2}} - x_{n+\frac{3N}{4}}$, given $((nk))_N = p$. From (12), the distance between the two successive data elements in a positive or negative group is given by $\frac{N}{g(k,N)}$. Since k is a divisor of N, this reduces to $\frac{N}{k}$. Using Theorem 3 (b), if elements $x_{n'}$, and x_n co-occur in $Y_k^{(p)}$ but having different signs, then,

$$n' = n + q_{odd}\frac{N}{2k} \tag{54}$$

where $q_{odd}$is an odd integer. At the next higher frequency $k' = 2k$, Eq. (54) becomes

$$n' = n + q_{odd}\frac{N}{k'} \tag{55}$$

From (12), the general form for a data element $x_{n'}$ of same sign present along with an element x_n in the MRT coefficient $Y_{k'}^{(p)}$ is given by (since k' is a divisor of N, $g(k', N) = k'$)

$$n' = n + j\frac{N}{k'} \tag{56}$$

where $j = 0, 1, 2, 3, \ldots, k' - 1$.

Equation (55) can be seen to be a special case of Eq. (56). For any higher frequency $k' = 2^k$, Eq. (56) holds. Hence, given k is the lowest frequency at which $x_{n'}$ and x_n occur with opposite signs, for all larger frequencies $k' = 2^k$, $x_{n'}$ and x_n occur with similar signs. For $N = 8$, $Y_1^{(0)} = x_0 - x_4$, $Y_2^{(0)} = x_0 - x_2 + x_4 - x_6$ and $Y_4^{(0)} = x_0 - x_1 + x_2 - x_3 + x_4 - x_5 + x_6 - x_7$. Data elements x_4 and x_0 occur with opposite signs in $Y_1^{(0)}$ and similar signs in higher frequencies, $k = 2, k = 4$. Conversely, another conclusion is that elements $x_{n'}$ and x_n that co-occur with same signs in a UMRT coefficient having frequency k' also co-occur with opposite signs in a UMRT coefficient having frequency k where $k' = 2^k$.

Given k is the lowest frequency where $x_{n'}$ and x_n co-occur with an opposite sign, the factor of multiplication for $x_{n'}$ in the inverse transform formula is $-\frac{1}{2k}$. In the case of higher frequencies till M, the factor of multiplication is $\frac{1}{2k'}, k' = 2k, 4k, \ldots M$. Thus the sum of the series

$$f = \frac{1}{N} - \frac{1}{2k} + \frac{1}{4k} + \frac{1}{8k} + \ldots \frac{1}{2M} \tag{57}$$

will provide the value of the multiplication factor f associated with element $x_{n'}$. Assume $k = \frac{N}{q}$. First the sum of the following series can be found:

$$\frac{1}{4k} + \frac{1}{8k} + \ldots \frac{1}{2M} = f_a$$

$$f_a = \sum_{j=\log_2 \frac{2N}{q}}^{\log_2 M} \frac{1}{2^{j+1}}$$

The number of terms in this summation is $\log_2(q) - 1$.

$$f_a = \frac{q}{4N}\frac{1 - 2^{-(\log_2(q)-1)}}{\frac{1}{2}} = \frac{q}{2N}\frac{q-2}{q} = \frac{q-2}{2N}$$

From Eq. (57),

$$f = \frac{1}{N} - \frac{1}{2k} + f_a = \frac{1}{N} - \frac{1}{2k} + \frac{q-2}{2N} = \frac{q}{2N} - \frac{1}{2k} = 0$$

Thus, all the other data elements $x_{n'}$ that occur along with x_n in the various MRT coefficients in the summation of the inverse formula cancel out, leaving behind only the desired data element x_n.

Hence, the formula for inverse UMRT is proved.

5.2 Theorem 8: N not a power of 2

Given the UMRT of a 1-D signal of size N, N not being a power of 2, the 1-D signal can be reconstructed from its UMRT by the following formula

$$x_n = \frac{1}{k'} Y_{k'_{((nk')_N)}} + \sum_{t=0}^{\log_2 k' - 1} \frac{1}{2^{t+1}} Y_{2^t}^{((2^t n)_N)} \tag{58}$$

where k' is the highest frequency index that is a power of 2 and also a divisor of N.

Proof. x_n has to be obtained from the UMRT. Given $Y_k^{(p)}$ data element x_n has to satisfy $((nk))_N = p$. For $k = 2^a$, the UMRT coefficient $Y_{2^a}^{((2^a n)_N)}$ contains x_n. The UMRT coefficient $Y_{k'_{((nk')_N)}}$ is multiplied by $\frac{1}{k'}$, and the other UMRT coefficients are multiplied by $\frac{1}{2^{t+1}}$. The resultant factor f that multiplies x_n as a result of the summation can be proved by using the same method used earlier. Similarly, it can also be shown that data elements other than x_n cancel out in the summation, leaving behind only the desired data element x_n. Hence, the proposed formula is proved.

5.3 General formula for any even N

Equation (58) can be generalized to be applicable to any even value of N. Hence, the following equation can be used for signal reconstruction from UMRT, for a signal of size N, N being any even number.

$$x_n = \frac{1}{k'} Y_{(k')_{N((nk')_N)}} + \sum_{t=0}^{\log_2 k' - 1} \frac{1}{2^{t+1}} Y_{2^t}^{((2^t n)_N)} \tag{59}$$

where k' is the highest power of 2 divisor of N.

For N a power of 2, the UMRT basis matrix can be defined in a new form by combining the frequency index k and the phase index p of an MRT coefficient into an index q as follows:

$$H_0(m) = H_{0,0}(m) = 1$$
$$1, \ [nk]_N = p$$

$$H_q(m) = H_{k,p}(m) = \begin{cases} -1, & [nk]_N = p + \frac{N}{2} \\ 0, & otherwise \end{cases}$$

$$q = 2^k + p - 1$$

$$m = 0, 1, \ldots, N - 1$$

This structure of the UMRT definition has a similar form as that of the Haar transforms [10]. **Table** 3 shows the mapping for $N = 8$.

k,p	0,0	1,0	1,1	1,2	1,3	2,0	2,2	4,0
q	0	1	2	3	4	5	6	7

Table 3.
Proposed mapping between 1-D UMRT indices and array indices, for N = 8.

6. Conclusions and future development

The MRT is a new representation of signals and involves only real additions. However, the MRT is expansive and redundant. The UMRT removes these features of MRT to give a real, invertible, nonexpansive signal transform for any even values of N. The MRT belongs in the same family of transforms as the Haar transform and the Hadamard transform. A number-theoretical foundation has hereby been laid for the 1-D MRT. However, the 2-D version of the MRT also exists. The interconnections between the 1-D version and 2-D version need to be studied further. The real merit of this transform is its utmost simplicity in terms of computational requirements. Simple integer-to-integer transforms will retain their attractiveness in the light of the ongoing switch to Internet of Things (IoT) and edge computing. A potential application of the MRT is its use as a feature vector in various contexts, just like how the Hadamard Transform has been used, as shown in [18–20]. Numerous other applications await the MRT.

Acknowledgements

The author would like to acknowledge the guidance and contributions of Dr. R. Gopikakumari.

Author details

Rajesh Cherian Roy
Muthoot Institute of Technology and Science, Ernakulam, Kerala, India

*Address all correspondence to: rajeshcherian@yahoo.co.in

References

[1] Ahmed N et al. Discrete cosine transform. IEEE Transactions on Computers. 1974;**C-23**:90-93. DOI: 10.1109/T-C.1974.223784

[2] Bracewell RN. Discrete Hartley transform. Journal of the Optical Society of America. 1983;**73**:1832-1835. DOI: 10.1364/JOSA.73.001832

[3] Ersoy OK. A two-stage representation of DFT and its applications. IEEE Transactions on Acoustics, Speech, and Signal Processing. 1987;**35**:825-831. DOI: 10.1109/TASSP.1987.1165202

[4] Roy RC, Gopikakumari R. A new transform for 2-d signal representation (MRT) and some of its properties. In: Proceedings of the International Conference in Signal Processing and Communication (SPCOM '04). Bangalore: IEEE; 2004. pp. 363-367. DOI: 10.1109/SPCOM.2004.1458423

[5] Lun DPK, Hsung TC, Shen TW. Orthogonal discrete periodic Radon transform. Part I: Theory and realization. Signal Processing. 2003;**83**: 941-955. DOI: 10.1016/S0165-1684(02)00498-X

[6] Roy RC, Anish Kumar MS, Gopikakumari R. An invertible transform for image representation and its application to image compression. In: Proceedings of the International Symposium in Signal Processing (ISSPA '07). Sharjah: IEEE; 2007. DOI: 10.1109/ISSPA.2007.4555504

[7] Anish Kumar MS, Roy RC, Gopikakumari R. A new transform coder for gray scale images using 4×4 MRT. Elsevier AEU—International Journal of Electronics and Communications. 2008;**62–8**:627-630. DOI: 10.1016/j.aeue.2007.04.009

[8] Roy RC, Gopikakumari R. 1-D MRT: A new transform for 1-D signals. In: Proceedings of the all India Seminar on Innovative Electronics Technology and Futuristic Communication, Kochi. 2009. pp. 85-88

[9] Fino J. Relations between Haar and Walsh/Hadamard transforms. Proceedings of the IEEE. 1972;**60**: 647-648. DOI: 10.1109/PROC.1972.8719

[10] Roy RC. Relationship between the Haar transform and the MRT. In: Proceedings of the Eighth International Conference on Information Communication and Signal Processing (ICICS '11), December 2011. Singapore: IEEE; 2011. DOI: 10.1109/ICICS.2011.6173127

[11] Roy RC. The relationship between the Hadamard transform and the MRT. In: Proceedings of the National Conference on Power Instrumentation Control and Computing (PICC '10), Thrissur. 2010

[12] Pei S, Chang K. Integer 2-D discrete Fourier transform pairs and eigenvectors using Ramanujan's sum. IEEE Signal Processing Letters. 2016;**23–1**:70-74. DOI: 10.1109/LSP.2015.2501421

[13] Qiu Q, Thompson A, Calderbank R, Sapiro G. Data representation using the Weyl transform. IEEE Transactions on Signal Processing. 2016;**64**(7): 1844-1853. DOI: 10.1109/TSP.2015.2505661

[14] Kamisli F. Lossless image and intra-frame compression with integer-to-integer DST. IEEE Transactions on Circuits and Systems for Video Technology. 2019;**29**(2):502-516. DOI: 10.1109/TCSVT.2017.2787638

[15] Roy RC. Development of a new transform: MRT [thesis]. Cochin University of Science and Technology; 2010

[16] Dence JB, Dence TP. Elements of the Theory of Numbers. USA: Academic Press; 1999

[17] Apostol TM. Introduction to Analytic Number Theory. USA: Springer; 1976

[18] Lakshmi PG, Dominic S. Walsh–Hadamard transform kernel-based feature vector for shot boundary detection. IEEE Transactions on Image Processing. 2014;**23**(12):5187-5197. DOI: 10.1109/TIP.2014.2362652

[19] Tang M, Chen X, Wen J, Han Y. Hadamard transform-based optimized HEVC video coding. IEEE Transactions on Circuits and Systems for Video Technology. 2019;**29**(3):827-839. DOI: 10.1109/TCSVT.2018.2810324

[20] Rout DK, Subudhi BN, Veerakumar T, Chaudhury S. Walsh–Hadamard-kernel-based features in particle filter framework for underwater object tracking. IEEE Transactions on Industrial Informatics. 2020;**16**(9):5712-5722. DOI: 10.1109/TII.2019.2937902

Chapter 2

Digit Sums and Infinite Products

Samin Riasat

Abstract

Consider the sequence u_n defined as follows: $u_n = +1$ if the sum of the base b digits of n is even, and $u_n = -1$ otherwise, where we take $b = 2$. Recall that the Woods-Robbins infinite product involves a rational function in n and the sequence u_n. Although several generalizations of the Woods-Robbins product are known in the literature, no other infinite product involving a rational function in n and the sequence u_n was known in closed form until recently. In this chapter we introduce a systematic approach to these products, which may be generalized to other values of b. We illustrate the approach by evaluating a large class of similar infinite products.

Keywords: radix representations, digit sums, Prouhet-Thue-Morse sequence, Woods and Robbins product, closed formulas for infinite products

1. Introduction

Throughout this chapter n will denote a non-negative integer. Let $s_b(n)$ denote the sum of the base b digits of n, and put $u_n = (-1)^{s_2(n)}$. We study infinite products of the form

$$f(b,c) := \prod_{n=1}^{\infty} \left(\frac{n+b}{n+c}\right)^{u_n}. \tag{1}$$

(We show in Section 2 that $f(b,c)$ converges for $b, c \in \mathbb{C} \setminus \{-1, -2, -3, \ldots\}$).

Plainly $f(b,c) = 1/f(c,b)$ and $f(b,b) = 1$. Up to these relations, it seems that the only known nontrivial value of f is $f(1/2, 1) = \sqrt{2}$, which is the famous Woods-Robbins identity [1, 2]:

$$\prod_{n=0}^{\infty} \left(\frac{2n+1}{2n+2}\right)^{u_n} = \frac{1}{\sqrt{2}}. \tag{2}$$

Several infinite products inspired by it were discovered afterwards (see, e.g., [3, 4]). But none of them involve the sequence u_n. Moreover, almost nothing is known (see, e.g., [5, 6]) about the similar product

$$\prod_{n=1}^{\infty} \left(\frac{2n}{2n+1}\right)^{u_n} = f\left(0, \frac{1}{2}\right). \tag{3}$$

Our goal is to study these infinite products in detail. This will allow us to gain a deeper understanding of such products as well as evaluate more products like the Woods-Robbins identity.

The material in this chapter is based on the two papers [7, 8].

2. General properties of the function f

First we establish a general result from [7] on convergence.

Lemma 1.1 Let $R \in \mathbb{C}(X)$ be a rational function such that the values $R(n)$ are defined and nonzero for $n \geq 1$. Then, the infinite product $\prod_n R(n)^{u_n}$ converges if and only if the numerator and the denominator of R have the same degree and same leading coefficient.

Proof. If the infinite product converges, then $R(n)$ must tend to 1 when n tends to infinity. Thus the numerator and the denominator of R have the same degree and the same leading coefficient.

Now suppose that the numerator and the denominator of R have the same leading coefficient and the same degree. Decomposing them in factors of degree 1, it suffices, for proving that the infinite product converges, to show that infinite products of the form

$$\prod_{n=1}^{\infty} \left(\frac{n+b}{n+c} \right)^{u_n} \tag{4}$$

converge for complex numbers b and c such that $n+b$ and $n+c$ do not vanish for any $n \geq 1$. Since the general factor of such a product tends to 1, it is equivalent, grouping the factors pairwise, to proving that the product

$$\prod_{n=1}^{\infty} \left[\left(\frac{2n+b}{2n+c} \right)^{u_{2n}} \left(\frac{2n+1+b}{2n+1+c} \right)^{u_{2n+1}} \right] \tag{5}$$

converges. Since $u_{2n} = u_n$ and $u_{2n+1} = -u_n$, we only need to prove that the infinite product

$$\prod_{n=1}^{\infty} \left(\frac{(2n+b)(2n+1+c)}{(2n+c)(2n+1+b)} \right)^{u_n} \tag{6}$$

converges. Taking (the principal determination of) logarithms, we see that

$$\log \left(\frac{(2n+b)(2n+1+c)}{(2n+c)(2n+1+b)} \right) = O\left(\frac{1}{n^2} \right) \tag{7}$$

which gives the convergence result.

Hence $f(b,c)$ converges for any $b, c \in \mathbb{C} \backslash \{-1, -2, -3, \ldots\}$. Using the definition of u_n, it follows that for any $b, c, d \in \mathbb{C} \backslash \{-1, -2, -3, \ldots\}$,

1. $f(b,b) = 1$.

2. $f(b,c) f(c,d) = f(b,d)$.

3. $f(b,c) = \frac{c+1}{b+1} f\left(\frac{b}{2}, \frac{c}{2}\right) f\left(\frac{c+1}{2}, \frac{b+1}{2}\right)$.

One can ask the natural question: is f the unique function satisfying these properties?

2.1 A new function

Properties 1 and 2 above give

$$f(b,c) f(d,e) = \frac{f(b,c) f(c,d) f(d,e) f(d,c)}{f(c,d) f(d,c)} = \frac{f(b,e) f(d,c)}{f(c,c)} = f(b,e) f(d,c). \tag{8}$$

Hence we can rewrite property 3 as

$$f(b,c) = \frac{f\left(\frac{b}{2}, \frac{b+1}{2}\right)}{b+1} \Big/ \frac{f\left(\frac{c}{2}, \frac{c+1}{2}\right)}{c+1}. \tag{9}$$

Thus $f(b,c)$ can be computed using only the quantities $h(x) = f\left(\frac{x}{2}, \frac{x+1}{2}\right)$, via

$$f(b,c) = \frac{c+1}{b+1} \cdot \frac{h(b)}{h(c)}. \tag{10}$$

So understanding f is equivalent to understanding h, in the sense that each can be completely evaluated in terms of the other.

Taking $c = b + \frac{1}{2}$ in Eq. (10) gives the functional equation:

$$h(b) = \frac{b+1}{b+\frac{3}{2}} h\left(b + \frac{1}{2}\right) h(2b). \tag{11}$$

Similar questions can be asked for h: is it the unique solution to Eq. (11)? What about monotonic/continuous/smooth solutions?

3. Infinite products

3.1 Direct approach

Theorem 1.1 The following relations hold.

1. For $b, c \in \mathbb{C} \backslash \{-1, -2, -3, \ldots\}$,

$$\prod_{n=1}^{\infty} \left(\frac{(n+b)\left(n + \frac{b+1}{2}\right)\left(n + \frac{c}{2}\right)}{(n+c)\left(n + \frac{c+1}{2}\right)\left(n + \frac{b}{2}\right)} \right)^{u_n} = \frac{c+1}{b+1}. \tag{12}$$

2. For $b, c \in \mathbb{C} \backslash \{0, -1, -2, \ldots\}$,

$$\prod_{n=0}^{\infty} \left(\frac{(n+b)\left(n + \frac{b+1}{2}\right)\left(n + \frac{c}{2}\right)}{(n+c)\left(n + \frac{c+1}{2}\right)\left(n + \frac{b}{2}\right)} \right)^{u_n} = 1. \tag{13}$$

3. For $b \in \mathbb{C} \backslash \{-1, -2, -3, \ldots\}$,

$$\prod_{n=1}^{\infty} \left(\frac{(n+b)\left(n + \frac{b+1}{4}\right)}{\left(n + \frac{b+3}{4}\right)\left(n + \frac{b}{2}\right)} \right)^{u_n} = \frac{b+3}{2(b+1)}. \tag{14}$$

4. For $c \in \mathbb{C} \backslash \{-1, -2, -3, \ldots\}$,

$$\prod_{n=1}^{\infty} \left(\frac{\left(n + \frac{1}{2}\right)\left(n + \frac{c}{2}\right)}{(n+c)\left(n + \frac{c+1}{2}\right)} \right)^{u_n} = c+1. \tag{15}$$

Proof.

1. This follows immediately using properties 1–3 in Section 2.

2. As above.

3. Take $c = (b+1)/2$ in Eq. (12).

4. Take $b = 0$ in Eq. (12).

Corollary 1.1 For any positive rational number q, there exist monic polynomials $P, Q \in \frac{1}{4}\mathbb{Z}[X]$, both at most cubic, such that

$$\prod_{n=1}^{\infty} \left(\frac{P(n)}{Q(n)}\right)^{u_n} = q. \tag{16}$$

Furthermore, if q is an integer, then P and Q can be chosen to be at most quadratic.

We still do not know exactly which numbers are given by such infinite products.

3.2 Functional equation approach

Recall the functional Eq. (11):

$$h(b) = \frac{b+1}{b+\frac{3}{2}} h\left(b+\frac{1}{2}\right) h(2b). \tag{17}$$

Taking $b = 0$ in Eq. (11) gives

$$h(0) = \frac{2}{3} h\left(\frac{1}{2}\right) h(0), \tag{18}$$

i.e., $h(1/2) = 3/2$. This shows that

$$\prod_{n=0}^{\infty} \left(\frac{4n+3}{4n+1}\right)^{u_n} = 2. \tag{19}$$

Next, taking $b = 1/2$ in Eq. (11) gives

$$h\left(\frac{1}{2}\right) = \frac{3}{4} h(1)^2; \tag{20}$$

hence $h(1) = \sqrt{2}$, and we recover the Woods-Robbins identity

$$\prod_{n=0}^{\infty} \left(\frac{2n+2}{2n+1}\right)^{u_n} = \sqrt{2}. \tag{21}$$

Similarly, taking $b = -1/2$ in Eq. (11) gives

$$h\left(-\frac{1}{2}\right) = \frac{1}{2} h(0) h(-1) = \frac{1}{2} f\left(0, \frac{1}{2}\right) f\left(-\frac{1}{2}, 0\right) = \frac{1}{2} f\left(-\frac{1}{2}, \frac{1}{2}\right), \tag{22}$$

i.e.,

$$\prod_{n=1}^{\infty} \left(\frac{(4n-1)(2n+1)}{(4n+1)(2n-1)}\right)^{u_n} = \frac{1}{2}. \tag{23}$$

Taking $b = 1$ in Eq. (11) gives

$$h(1) = \frac{4}{5} h\left(\frac{3}{2}\right) h(2); \tag{24}$$

hence $h(3/2)h(2) = 5\sqrt{2}/4$, and this gives

$$\prod_{n=0}^{\infty} \left(\frac{(4n+3)(2n+2)}{(4n+5)(2n+3)}\right)^{u_n} = \frac{1}{\sqrt{2}}. \tag{25}$$

Taking $b = 3/2$ in Eq. (11) and using the previous result gives

$$h(2)^2 h(3) = \frac{3}{\sqrt{2}} \tag{26}$$

which is equivalent to

$$\prod_{n=0}^{\infty} \left(\frac{(2n+2)(n+1)}{(2n+3)(n+2)}\right)^{u_n} = \frac{1}{\sqrt{2}}. \tag{27}$$

These identities can be also combined in pairs to obtain other identities.

4. Some analytical results

We saw in the previous section that some of the infinite products we evaluated were integers, some were rational, and some were quadratic irrational. In the hope of further understanding their nature, we now study the analytical behaviors of f and h.

Lemma 1.2 Let $b, c \in (-1, \infty)$.

1. If $b = c$, then $f(b,c) = 1$.

2. If $b > c$, then

$$\left(\frac{c+1}{b+1}\right)^2 < f(b,c) < 1. \tag{28}$$

3. If $b < c$, then

$$1 < f(b,c) < \left(\frac{c+1}{b+1}\right)^2. \tag{29}$$

Proof. Using properties 1–3 from Section 2, it suffices to prove 2.
Let $b > c > -1$, and put

$$a_n = \log\left(\frac{n+b}{n+c}\right), \quad S_N = \sum_{n=1}^{N} a_n u_n, \quad U_N = \sum_{n=1}^{N} u_n. \tag{30}$$

Note that a_n is positive and strictly decreasing to 0. Using $s_2(2n+1) = 1 + s_2(2n)$, it follows that $U_n \in \{-2, -1, 0\}$ and $U_n \equiv n (\mathrm{mod} 2)$, for each n. Using summation by parts,

$$S_N = a_{N+1}U_N + \sum_{n=1}^{N} U_n(a_n - a_{n+1}). \tag{31}$$

So $-2a_1 < S_N < 0$ for $N \geq 2$. Exponentiating and taking $N \to \infty$ gives the desired result.

Lemma 1.2 together with Eq. (10) implies the following results.

Theorem 1.2 $h(x)/(x+1)$ is strictly decreasing on $(-1, \infty)$, and $h(x)(x+1)$ is strictly increasing on $(-1, \infty)$.

Proof. Let $-1 < b < c$. By Eqs. (29) and (10),

$$1 < \frac{c+1}{b+1} \cdot \frac{h(b)}{h(c)} < \left(\frac{c+1}{b+1}\right)^2 \tag{32}$$

from which the result follows.

Theorem 1.3 For $b, c \in (-1, \infty), f(b,c)$ is strictly decreasing in b and strictly increasing in c.

Proof. By Eq. (10),

$$\frac{h(c)f(b,c)}{c+1} = \frac{h(b)}{b+1} \tag{33}$$

and

$$\frac{(b+1)f(b,c)}{h(b)} = \frac{c+1}{h(c)} \tag{34}$$

hence the result follows from Theorem 1.2.

Theorem 1.4 For $x \in (-2, \infty)$,

$$1 < h(x) < \left(\frac{x+3}{x+2}\right)^2. \tag{35}$$

Proof. This follows from taking $b = x/2$ and $c = (x+1)/2$ in Eq. (10), then using Eq. (29).

We now prove some results on differentiability.

Theorem 1.5 $h(x)$ is smooth on $(-2, \infty)$.

Proof. Take $b = x/2$ and $c = (x+1)/2$ in Eq. (30). Then the sequence S_n of smooth functions on $(-2, \infty)$ converges pointwise to $\log h$.

Differentiating with respect to x gives

$$S'_N = \sum_{n=1}^{N} \frac{u_n}{(2n+x)(2n+1+x)} = \sum_{n=1}^{N} u_n \left(\frac{1}{2n+x} - \frac{1}{2n+1+x}\right). \tag{36}$$

Hence

$$\begin{aligned} |S'_N - S'_M| &\leq \sum_{n=M+1}^{N} \left(\frac{1}{2n+x} - \frac{1}{2n+1+x}\right) \\ &\leq \sum_{n=M+1}^{N} \left(\frac{1}{2n-1+x} - \frac{1}{2n+1+x}\right) \\ &= \frac{1}{2M+1+x} - \frac{1}{2N+1+x} \\ &< \frac{1}{2M-1} \to 0 \end{aligned} \tag{37}$$

as $M \to \infty$, for any $x \in (-2, \infty)$ and $N > M$. Thus S_n' converges uniformly on $(-2, \infty)$, which shows that $\log h$, hence h, is differentiable on $(-2, \infty)$.

Now suppose that derivatives of h up to order k exist for some $k \geq 1$. Note that

$$S_N^{(k+1)} = (-1)^k k! \sum_{n=1}^{N} u_n \left(\frac{1}{(2n+x)^{k+1}} - \frac{1}{(2n+1+x)^{k+1}} \right). \tag{38}$$

As before,

$$\begin{aligned} \left| S_N^{(k+1)} - S_M^{(k+1)} \right| &\leq k! \sum_{n=M+1}^{N} \left(\frac{1}{(2n+x)^{k+1}} - \frac{1}{(2n+1+x)^{k+1}} \right) \\ &\leq k! \sum_{n=M+1}^{N} \left(\frac{1}{(2n-1+x)^{k+1}} - \frac{1}{(2n+1+x)^{k+1}} \right) \\ &= \frac{k!}{(2M+1+x)^{k+1}} - \frac{k!}{(2N+1+x)^{k+1}} \\ &< \frac{k!}{(2M-1)^{k+1}} \to 0 \end{aligned} \tag{39}$$

as $M \to \infty$, for any $x \in (-2, \infty)$ and $N > M$. Hence $S_n^{(k+1)}$ converges uniformly on $(-2, \infty)$, i.e., $h^{(k)}$ is differentiable on $(-2, \infty)$.

Therefore, by induction, h has derivatives of all orders on $(-2, \infty)$.

Theorem 1.6 Let $a \geq 0$. Then

$$\log h(x) = \log h(a) + \sum_{k=1}^{\infty} \frac{(-1)^{k-1}}{k} \left(\sum_{n=2}^{\infty} \frac{u_n}{(n+a)^k} \right) (x-a)^k \tag{40}$$

for $x \in [a-1, a+1]$.

Proof. Let $H(x) = \log h(x)$. By Theorem 1.5,

$$H^{(k+1)}(x) = (-1)^k k! \sum_{n=2}^{\infty} \frac{u_n}{(n+x)^{k+1}}. \tag{41}$$

Hence

$$|H^{(k+1)}(x)| \leq k! \sum_{n=2}^{\infty} \frac{1}{|n+x|^{k+1}} \leq k! \sum_{n=2}^{\infty} \frac{1}{(n+a-1)^{k+1}} \tag{42}$$

for $x \in [a-1, a+1]$. So by Taylor's inequality, the remainder for the Taylor polynomial for $H(x)$ of degree k is absolutely bounded above by

$$\frac{1}{k+1} \left(\sum_{n=2}^{\infty} \frac{1}{(n+a-1)^{k+1}} \right) |x-a|^{k+1} \tag{43}$$

which tends to 0 as $k \to \infty$, since $a \geq 0$ and $|x-a| \leq 1$. Therefore $H(x)$ equals its Taylor expansion about a for x in the given range.

5. Further remarks on $h(0)$

As mentioned in Section 1, not much is known about the quantity $h(0) \approx 1.62816$. We give the following explanation as to why $h(0)$ might behave specially in a sense.

Note that the only way nontrivial cancelation occurs in Eq. (11) is when $b = 0$. Likewise, nontrivial cancelation occurs in Eq. (10) or property 3 in Section 2 only for $(b, c) = (0, 1/2)$ and $(1/2, 0)$. That is, the victim of any such cancelation is always $h(0)$ or $h(0)^{-1}$. So we must look for other ways to study $h(0)$.

Using the two known values $h(1/2) = 3/2$ and $h(1) = \sqrt{2}$, the following expressions for $h(0)$ are obtained from Theorem 1.6 by choosing various values for x and a.

- $x = 0$ and $a = 1$:

$$h(0) = \sqrt{2} \exp\left(-\sum_{k=1}^{\infty} \frac{1}{k} \sum_{n=2}^{\infty} \frac{u_n}{(n+1)^k}\right) \tag{44}$$

- $x = 1$ and $a = 0$:

$$h(0) = \sqrt{2} \exp\left(\sum_{k=1}^{\infty} \frac{(-1)^k}{k} \sum_{n=2}^{\infty} \frac{u_n}{n^k}\right) \tag{45}$$

- $x = 0$ and $a = 1/2$:

$$h(0) = \frac{3}{2} \exp\left(\sum_{k=1}^{\infty} \frac{1}{k} \sum_{n=2}^{\infty} \frac{u_{2n+1}}{(2n+1)^k}\right) \tag{46}$$

- $x = 1/2$ and $a = 0$:

$$h(0) = \frac{3}{2} \exp\left(\sum_{k=1}^{\infty} \frac{(-1)^k}{k} \sum_{n=2}^{\infty} \frac{u_{2n}}{(2n)^k}\right) \tag{47}$$

The Dirichlet series appearing in the above expressions were studied in [9]. We think that these identities and the results from Section 4 might help in shedding some light on the nature of $h(0)$.

6. Conclusions and future developments

We evaluated infinite products involving the digit sum function $s_b(n)$ by splitting the product based on the congruence classes modulo b. We illustrated two approaches for doing so, one by direct computation and another using functional equations. For $b = 2$ we proved some analytical results to aid us in understanding the behavior of these products. Many open questions still remain.

Although we only considered the base $b = 2$, many of the results above easily generalize to other bases. One possible direction toward a generalization is to take $u_n = (-1)^{s_b(n)}$. Another is $u_n = \omega_b^{s_b(n)}$, where ω_b is a primitive b-th root of unity. We leave these as work to be done in the future.

Author details

Samin Riasat
University of Waterloo, Waterloo, Canada

*Address all correspondence to: sriasat@uwaterloo.ca

References

[1] Robbins D. Solution to problem E 2692. American Mathematical Monthly. 1979;**86**:394-395

[2] Woods DR. Elementary problem proposal E 2692. American Mathematical Monthly. 1978;**85**:48

[3] Allouche J-P, Shallit J. Infinite products associated with counting blocks in binary strings. Journal of the London Mathematical Society. 1989;**39**: 193-204

[4] Allouche J-P, Sondow J. Infinite products with strongly B-multiplicative exponents. Annales Universitatis Scientiarum Budapestinensis de Rolando Eötvös Nominatae. Sectio Biologica/Errata: Annales Universitatis Scientiarum Budapestinensis de Rolando Eötvös Nominatae. Sectio Biologica. 2008/2010;**28/32**:35-53/253

[5] Allouche J-P. Thue, combinatorics on words, and conjectures inspired by the Thue-Morse sequence. Journal de Théorie des Nombres de Bordeaux. 2015;**27**(2):375-388

[6] Allouche J-P, Shallit J. The ubiquitous Prouhet-Thue-Morse sequence. Sequences and their Applications. In: Ding C, Helleseth T, Niederreiter H, editors. Proceedings of SETA'98. Springer Verlag. 1999; 1-16

[7] Allouche J-P, Riasat S, Shallit J. More infinite products: Thue-Morse and the gamma function. The Ramanujan Journal. 2019;**49**:115-128. DOI: 10.1007/s11139-017-9981-7

[8] Riasat S. Infinite products involving binary digit sums. In: Kilgour D, Kunze H, Makarov R, Melnik R, Wang X, editors. Recent Advances in Mathematical and Statistical Methods. AMMCS 2017. Springer Proceedings in Mathematics & Statistics. Vol. 259. Cham: Springer; 2017

[9] Allouche J-P, Cohen H. Dirichlet series and curious infinite products. The Bulletin of the London Mathematical Society. 1985;**17**:531-538

Chapter 3

The Borel-Cantelli Lemmas, and Their Relationship to Limit Superior and Limit Inferior of Sets (or, Can a Monkey Really Type Hamlet?)

Anant P. Godbole

Abstract

The purpose of this chapter is to show that if a monkey types infinitely, Shakespeare's *Hamlet* and any other works one may wish to add to the list will *each* be typed, not once, not twice, but *infinitely often* with a probability of 1. This dramatic fact is a simple consequence of the *Borel-Cantelli lemma* and will come as no surprise to anyone who has taken a graduate-level course in Probability. The proof of this result, however, is quite accessible to anyone who has but a rudimentary understanding of the concept of independence, together with the notion of limit superior and limit inferior of a sequence of sets.

Keywords: Borel-Cantelli lemma, limit superior of sets, limit imferior of sets, independence, limit theorems of probability

1. Introduction

Consider a monkey named Sue who is given a word processor with N symbols. We shall assume that these symbols include the 26 letters of the English alphabet (upper and lower case), all the Greek letters, the numbers 0 through 9, a blank space, all the standard punctuation marks (,.; – etc.), and mathematical symbols (∞, $\int$, $\Rightarrow$, ∇, etc.); imagine, in fact, that N is so large that the keyboard is capable of typing just anything we might fancy, in any language. (A LaTeX editor could do much of that too, but not in all languages!)

If Sue is handed such a machine and pounds away, randomly, it is clear that most of what she types will be complete gibberish. A stray segment of sense such as "dog eat HAy" or even "Ronnie ReAGan our 40th PresiD$\varepsilon\eta\tau$ of USA" would surprise no one, but how often, we ask, would she successfully manage to type the Constitution of the United States, or Shakespeare's *Hamlet*, or the fundamental mathematical works of the 2026 Fields Medalist(s)?

The purpose of this chapter is to show that if Sue types infinitely, the above works (and any others that one may choose to add to the list) will *each* be typed, not once, not twice, but *infinitely often* with a probability of 1. This dramatic fact is a simple consequence of the *Borel-Cantelli lemma* and will come as no surprise to anyone who has taken a serious graduate-level course in Probability. The proof of this result, however, is quite accessible to anyone who has but a rudimentary understanding of the concept of independence.

The reader is invited, while reading this chapter, to let his/her imagination run wild, and concoct a plethora of similar examples. A somewhat mundane objection may be raised immediately: how can Sue (or anyone else for that matter) type indefinitely? We shall not dwell on this nonmathematical problem, but will remark instead (and prove a little later) that Sue's never-ending assignment is mathematically equivalent to the task of randomly selecting a number from the interval $[0, 1]$.

We would like to mention that our problem is related to the famous "Problem of a printed line," a popular account of which can be found in George Gamow's classic book [1]. The solution presented there, however, is entirely deterministic and of a finite character: the automatic printing press considered by Gamow does not print indefinitely, and the probabilities of various outcomes are not calculated.

2. Limit superior and limit inferior of a sequence of sets

Consider any sequence $\{A_n\}_{n=1}^{\infty}$ of subsets of a set Ω. Points of Ω will be denoted by ω. We know that

$$\bigcup_{n=1}^{\infty} A_n = \{\omega \in \Omega : \omega \in A_n \text{ for some } n\}$$

and

$$\bigcap_{n=1}^{\infty} A_n = \{\omega \in \Omega : \omega \in A_n \text{ for each } n\}$$

It follows that

$$\bigcap_{n=1}^{\infty} \bigcup_{k=n}^{\infty} A_k = \{\omega \in \Omega : \forall n \exists k \geq n \text{ with } \omega \in A_k\}$$

To better understand this somewhat complicated set, we first let $n = 1$ and note that $\omega \in A_k$ for some $k \geq 1$, say $k = k_1$. Letting $n = k_1$, we see that ω must belong to some A_{k_2}, where $k_2 \geq k_1$. Continuing in this fashion, we see that $\omega \in \cap_{n=1}^{\infty} \cup_{k=n}^{\infty} A_k$ if and only if $\omega \in A_k$ for infinitely many k's.

The set $\cap_{n=1}^{\infty} \cup_{k=n}^{\infty} A_k$ is called the *limit superior* of the sequence $\{A_n\}_{n=1}^{\infty}$, and is denoted by $\limsup A_n$, or, $\overline{\lim} A_n$, or, rather appropriately, by $\{A_n \text{ i.o.}\}$, where i.o. stands for "infinitely often."

In a similar fashion, we observe that the set

$$\bigcup_{n=1}^{\infty} \bigcap_{k=n}^{\infty} A_k = \{\omega \in \Omega : \exists n \text{ such that } \omega \in A_k \ \forall k \geq n\}$$

is a collection of those points ω that belong to *all but a finite number* of the A_n's. $\cup_{n=1}^{\infty} \cap_{k=n}^{\infty} A_k$ is called the *limit inferior* of the sequence $\{A_n\}_{n=1}^{\infty}$ and is usually denoted by $\liminf A_n$ or $\underline{\lim} A_n$. We prefer the notation $\{A_n \text{ a.b.f.o}\}$ (a.b.f.o means "all but finitely often"). Elementary symbol manipulation may be used to prove that $\underline{\lim} A_n \subset \overline{\lim} A_n$. It is easier to note, however, that if ω belongs to all but finitely many A_n's it must necessarily belong to an infinite number of them. The above fact is just one of the many similarities between lim sups and lim infs of sets, on the one hand, and of real numbers, on the other (recall that $\lim\ \inf_{n\to\infty} a_n \le \lim\ \sup_{n\to\infty} a_n$). Likewise $\lim\inf_{n\to\infty} a_n$ and $\lim\ \sup_{n\to\infty} a_n$ must both exist, that is,

$$-\infty \le \lim\ \inf a_n \le \lim\ \sup a_n \le \infty,$$

as must $\underline{\lim} A_n$ and $\overline{\lim} A_n$, with

$$\phi \subset \underline{\lim} A_n \subset \overline{\lim} A_n \subset \Omega.$$

If $\{A_n \text{ a.b.f.o}\} = \{A_n \text{ i.o.}\}$, the sequence $\{A_n\}$ is said to have a limit, which we define to be the common value of $\{A_n \text{ a.b.f.o}\}$ and $\{A_n \text{ i.o.}\}$ and denote by $\lim_{n\to\infty} A_n$ (note, again, the analogy with real sequences).

A useful dual relation between these two sets is

$$\left(\overline{\lim} A_n\right)^c = \underline{\lim} A_n^c,$$

and

$$\left(\underline{\lim} A_n\right)^c = \overline{\lim} A_n^c$$

where A^c denotes the complement of the set A. Here is an informal proof of the first of these two facts: $\omega \in \underline{\lim} A_n^c$ iff ω belongs to all but finitely many A_n^c's iff $\omega \in$ just a finite number of the A_n's iff $\omega \notin \overline{\lim} A_n$ iff $\omega \in \left(\overline{\lim} A_n\right)^c$.

A few examples should help familiarize the reader with the above notions: the second and the third are taken from [2]:

Example 1. If $A_1 \subset A_2 \subset \dots$, then $\overline{\lim} A_n = \underline{\lim} A_n = \cup_{n=1}^{\infty} A_n$. Likewise, if $A_1 \supset A_2 \supset A_3 \supset \dots$, then $\overline{\lim} A_n = \underline{\lim} A_n = \cap_{n=1}^{\infty} A_n$.

Example 2. If $A_{2n-1} = \left(-\frac{1}{n}, 1\right]$ and $A_{2n} = \left(-1, \frac{1}{n}\right]$, then $\underline{\lim} A_n = \{0\}$ and $\overline{\lim} A_n = (-1, 1]$.

Example 3. if A_n is the unit circle with center at $\left(\frac{(-1)^n}{n}, 0\right)$, then

$$\underline{\lim} A_n = \left\{(x, y) : x^2 + y^2 < 1\right\}$$

and

$$\overline{\lim} A_n = \left\{(x, y) : x^2 + y^2 \le 1\right\} \backslash \{(0, 1), (0, -1)\}.$$

In what follows, the set Ω will be taken to be the *sample space* (or set of possible realizations) of a random experiment (one whose outcome cannot be predicted in advance). We shall assume that each subset of Ω that we encounter is *measurable*. In other words, each set A will be assumed to belong to the *sigma algebra* $\mathcal{A}$ of *events*, which is a class of subsets of Ω satisfying the conditions

- $\phi \in \mathcal{A}$
- If $A \in \mathcal{A}$, then $A^c \in \mathcal{A}$ and
- If $A_1, A_2, \dots \in \mathcal{A}$, then $\cup_{n=1}^{\infty} A_n \in \mathcal{A}$.

This restriction ensures that sets such as $\overline{\lim} A_n$ and $\underline{\lim} A_n$ are themselves measurable, so that we may meaningfully talk of their probabilities $P\{A_n \text{ i.o.}\}$ and $P\{A_n \text{ a.b.f.o}\}$. We next move on to a key concept in probability:

Definition: The sequence of events $\{A_n\}_{n=1}^{\infty}$ will be said to be *independent*, if for each finite subcollection $A_{n_1}, A_{n_2}, \ldots, A_{n_k}$,

$$P(A_{n_1} \cap A_{n_2} \cap \ldots \cap A_{n_k}) = P(A_{n_1}) \cdot P(A_{n_2}) \cdot \ \ldots \ \cdot P(A_{n_k}).$$

Stated informally, this means that the occurrence (or nonoccurrence) of any finite subcollection $\{A_{n_1}, A_{n_2}, \ldots, A_{n_k}\}$ does not affect the probability of occurrence of another disjoint collection $\{A_{m_1}, A_{m_2}, \ldots, A_{m_\ell}\}$.

The events $\{A_n\}_{n=1}^{\infty}$ that represent the successive outcomes of an infinite coin-tossing experiment are usually assumed, on intuitive and empirical grounds, to be independent. We shall make the same assumption regarding Sue's successive choices $\{B_n\}_{n=1}^{\infty}$ of a keyboard's key.

The *Borel-Cantelli lemma* is a two-pronged theorem, which asserts that the probability of occurrence of an infinite number of the independent events $\{A_n\}_{n=1}^{\infty}$ is zero or one:

Theorem 2.1. *(The Borel-Cantelli lemma,* [3, 4]*).*

a. If $\{A_n\}_{n=1}^{\infty}$ *is any sequence of events, then* $\sum_{n=1}^{\infty} P(A_n) < \infty$ *implies that* $P(A_n \text{ i.o.}) = 0$.

b. *If* $\{A_n\}_{n=1}^{\infty}$ *is an independence sequence, then* $P(A_n \text{ i.o.})$ *equals* 0 *or* 1 *according as the series* $\sum_{n=1}^{\infty} P(A_n)$ *converges or diverges.*

The following lemma can be proved using elementary properties of probability measures:

Lemma 2.2. *If* $\{A_n\}_{n=1}^{\infty}$ *is an increasing (decreasing) sequence of events then* $\lim_{n\to\infty} P(A_n) = P(\cup_{n=1}^{\infty} A_n)\,[P\cap_{n=1}^{\infty} A_n]$.

Proof of Theorem 2.1

a. For any n, we note that

$$P(A_n \text{ i.o.}) \le P(\cup_{k\ge n} A_k) \le \sum_{k=n}^{\infty} P(A_k)$$

On letting $n \to \infty$, we see that

$$P(A_n \text{ i.o.}) \le \lim_{n\to\infty} \sum_{k=n}^{\infty} P(A_k) = 0,$$

proving part (a).

b. We shall prove that $P(A_n^c \text{ a.b.f.o}) = 0$. Let $m, n \ (m \ge n)$ be arbitrary. Note that

$$P(\cap_{k=n}^{m} A_k^c) = \prod_{k=n}^{m} P(A_k^c) = \prod_{k=n}^{m} (1 - P(A_k)) \le \exp\left\{-\sum_{k=n}^{m} P(A_k)\right\},$$

where the last inequality follows from the fact that $1 - x \le e^{-x} \ (x \ge 0)$. Lemma 2.1 now gives us that

$$\lim_{m\to\infty} P\left(\bigcap_{k=n}^{m} A_k^c\right) = P\left(\bigcap_{k=n}^{\infty} A_k^c\right) = 0 \ \forall n,$$

which proves that

$$P(A_n^c \text{ a.b.f.o}) \le \sum_{n=1}^{\infty} P\left(\bigcap_{k=n}^{\infty} A_K^c\right) = 0,$$

proving the result.

As seen by the dates on Refs. [3, 4], the Borel-Cantelli lemmas are classical, and now part of virtually all graduate level books on Probability such as [2]. Since then, for over 100 years, the literature on the lemmas has focused on weakening the independence requirement in the second lemma, or looking at more complicated probability models that yield the same conclusions. See for example [5–7]. What distinguishes this work from these and others is that we provide a very down-to-earth application that forces the reader to come to terms with the notions of independence and infinity, as opposed to the finite samples one has in statistical situations. It is a paper that we feel can cause amusement, astonishment, false disbelief, and, ultimately, understanding. With this backdrop, we are now in a position to start establishing the claim made at the beginning of this chapter:

Corollary 2.3. *If Sue types indefinitely, by successively and independently choosing one of the N available keys, then any specific work containing a total of M characters (including all the blanks, of course) will be typed by her indfinitely often, with a probability of 1.*

Proof. Let A_1 be the event that Sue's first M random choices lead to the work being typed correctly. It is clear that $P(A_1) = \left(\frac{1}{N^M}\right)$. Similarly, let A_2 denote a successful completion of the task between the $(M+1)^{\text{st}}$ and $(2M)^{\text{th}}$ keystrokes. In general, A_n denotes the completion of a flawless job between $((n-1)M+1)^{\text{st}}$ and the $(nM)^{\text{th}}$ random strokes. It is evident that $P(A_n) = \left(\frac{1}{N^M}\right)$ $(n \ge 1)$ and that the sequence $\{A_n\}_{n=1}^{\infty}$ is independent. Since $\sum_{n=1}^{\infty} P(A_n) = \sum_{n=1}^{\infty}\left(\frac{1}{N^M}\right) = \infty$, it follows by part (b) of the Borel-Cantelli lemma that $P(A_n \text{ i.o.}) = 1$. Since the probability that the work is typed correctly infinitely often is at least as large as $P(A_n \text{ i.o.})$, the proof is complete. (Notice how the events $\{A_n\}_{n=1}^{\infty}$ are defined using disjoint blocks; this guarantees their independence.)

Lemma 2.4. *If* $P(C_n) = 1$ $(n = 1,2,\ldots)$, *then* $P\left(\cap_{n=1}^{\infty} C_n\right) = 1$.

Proof: Boole's inequality states that

$$P\left(\cap_{k=1}^{n} C_k\right) \ge \sum_{k=1}^{n} P(C_k) - (n-1).$$

Thus $P\left(\cap_{k=1}^{n} C_k\right) = 1$ for each n. The required conclusion is obtained on letting $n \to \infty$, and using Lemma 2.2.

Corollary 2.5. *If Sue types indefinitely, then every piece of writing (of any finite length whatsoever; published, unpublished, or yet to be written by yet unborn individuals; meaningful or gibberish) will be typed by her, infinitely often each, with a probability of 1.*

Proof. Denote the works by $B_1, B_2, \ldots$, and set $A_n = \{B_n \text{ i.o.}\}$, $(n = 1, 2, \ldots)$. By Lemma 2.2, $P(A_n) = 1$ for each n and thus, by Lemma 2.4, $P\left(\cap_{n=1}^{\infty} A_n\right) = 1$, as claimed. (Note: this implies that the probability that *some* work is typed finitely often (or never) is zero.)

Example 4. (*Statistical tests of hypotheses*) If a fair coin is tossed infinitely often, a sequence of 10^6 consecutive heads will appear infinitely often with probability 1.

Now, if a coin (of unknown origin) were tossed a million times, and a head appeared each time, the "null" statistical hypothesis

$$H_0 : \text{The coin is fair } (p = 1/2)$$

would be summarily rejected at most conventional (5, 1, 0.00001%) levels of significance. The point to note, however, is that such "extreme" and "erratic" behavior *will* be exhibited on an infinite number of occasions by any fair coin (and by all coins with $P(H) > 0$), with a probability of 1.

Similarly, if a fair coin is tossed infinitely often, an n-long alternating sequence $HTHT \ldots HT$ (n is arbitrary) will appear infinitely often, almost certainly. This fact may be compared with the conclusion of a standard nonparametric statistical procedure, *the run test*: the fair coin hypothesis would be vigorously rejected, using this test, if a large number of coin tosses yielded an alternating sequence of heads and tails.

3. A probability model for infinite coin tossing

In the above discussion, we often concluded that a particular event (e.g., *Hamlet* is typed infinitely often) occurred with probability 1. One fundamental question that we did not address, however, was the following: just what probability model describes infinite coin tossing or simian typewriting? Put another way, what are the sample spaces associated with these two experiments? And what exactly is the probability of an event defined to be? We realize then, in retrospect, that we had put the cart before the horse; various events were shown to have probability 1, by *assuming* the existence of a logically consistent probability (measure) on a sample space that had not been fully described. This practice is fairly standard in the teaching of probability; for example, sequences $\{X_n\}_{n=1}^{\infty}$ of independent and identically distributed (i.i.d.) random variables are often introduced as mathematical objects before their existence is proved, using Kolmogorov's famous Consistency Theorem. Such an approach is often beneficial; as Billingsley [8] wrote, "It is instructive... to see the show in rehearsal as well as in performance."

We shall start by noting that three tosses of a fair coin lead to the eight-point sample space

$$\Omega = \{HHH, HHT, HTH, HTT, THH, THT, TTH, TTT\}$$

It seems reasonable to assign probability 1/8 to each of these eight points; thus the probability $P(A)$ of *any* subset A may be defined by

$$P(A) = \frac{\text{number of points in } A}{8}$$

Our analysis is thus complete, and can easily be extended to any finite number of coin tosses. The situation gets rapidly more complicated if the coin is tossed endlessly. This experiment cannot be conceived, carried out, or justified "in practice," and our neat conclusions would be rendered meaningless if we were unable to mathematically model our procedure. Happily, however, this is not the case. We simply let

$$\begin{aligned} \Omega &= \{\omega : \omega \text{ is an infinite sequence of H's and T's}\} \\ &= \{\omega : \omega \text{ is an infinite sequence of 1's and 0's}\} \end{aligned}$$

A typical element of Ω might be $\omega = 0010011\ldots$. It is well known (and easily proved) that Ω is an uncountable set. It seems reasonable, then, to assign probability zero to each sample point.

The next step is crucial. We identify each element of Ω with the real number in the interval [0,1] that has the same binary expansion. For example, the sample outcome THTHTH ... is identified with the real number 0.01010101 ... which equals

$$\frac{0}{2^1}+\frac{1}{2^2}+\frac{0}{2^3}+\frac{1}{2^4}+\ldots=\frac{1}{3}.$$

A problem arises immediately: Numbers of the form $k/2^n$, where k and n are positive integers, do not have a unique binary representation. In other words, two different sample outcomes such as HTTTTT... and THHHH... would correspond to the same real number 1/2 (since 1/2 = 0.0111... = 0.100...), and the correspondence between Ω and [0,1] would not, consequently, be one-to-one. We note, however, that numbers of the form $k/2^n$ constitute a denumerable set, and that there are two sample outcomes that correspond to each such number. If one, but not both, of each of these outcomes were to be removed from Ω, we would be left with a one-to-one map from a censored sample space Ω' onto [0,1]. Moreover, our assumption regarding individual sample points forces $P(\Omega\backslash\Omega')$ to equal zero. Thus, if a set of zero probability is thrown out from the original sample space, we may let $\Omega = [0,1]$ and derive great satisfaction from the knowledge that this would not change the answer to any of our probability calculations.

It is possible to show, in a somewhat non-rigorous fashion (i.e., without using much measure theory), or rigorously, by introducing Lebesgue measure, that infinite coin tossing is mathematically equivalent to choosing a number randomly from the interval [0,1]. It can be shown, in a completely analogous way, that infinite random typewriting is equivalent to the single random choice of a number in [0,1]. We need of course, to consider the N-ary representation of numbers in [0,1], instead of their binary expansion (where N ithe number of typewriter keys). However, we shall not do so here.

Example 5. (*Random Numbers*) Let the random variable X denote the random choice of a number from [0,1]. Then

$$P(X\ \text{is}\ \text{rational})=\sum_{n=1}^{\infty}P(X=r_n,)$$

where r_n is the n'th rational. Since, $P(X=r_n)=0$ for each n, we have that

$$P(X\ \text{is}\ \text{rational})=0.$$

This result may be compared with a mundane fact of "reality": If a person, computer, pointer, or random number generator were asked to choose X, limitations of measurement accuracy (or decimal point restrictions) would systematically exclude irrational X's, leading to the "conclusion" that

$$P(X\ \text{is}\ \text{rational})=1!$$

We would like to next state a thrilling result, called *Borel's law of normal numbers* [3]: A number in [0,1] is said to be *normal*, if its decimal representation has, asymptotically, an equal frequency of the digits 0 through 9:

$$\lim_{n\to\infty}\left\{\frac{\text{the number of times } j \text{ appears in the first } n \text{ digits of its decimal expansion}}{n}\right\}=\frac{1}{10}$$

for each $j = 0, 1, 2, \ldots .9$. Borel's law states that

$$P(X \text{ is normal}) = 1,$$

which is somewhat surprising, since it is awfully hard to think of a single number that *is* normal (the number 0.012345678910111213..., obtained by writing each integer successively, is known to be normal; the proof is not trivial).

The Borel-Cantelli lemma yields several consequences that may, at first glance, seem to contradict Borel's normal number law:

Almost all the numbers in [0,1] (i.e., all except some with zero Lebesgue measure) have decimal expansions that contain infinitely many chains of length 1000, say, that contain no numbers except 2,3, and 4. The nice part is, of course, that almost all of these numbers are normal *as well*, and so on.

The moral of the Borel-Cantelli lemma should, by now, be quite clear: "The realization of a truly random infinite procedure will, with probability one, contain infinitely many segments that exhibit extreme 'non-randomness', of all sizes, patterns and intensities." The Borel-Cantelli lemma is, after all, a *limit theorem* of probability, and a quote from the classic treatise of Gnedenko and Kolmogorov [9] might be in order as well: "In reality, however, the epistemological value of the theory of probability is revealed only by limit theorems. Moreover, without limit theorems, it is impossible to understand the real content of the primary concept of all our sciences—the concept of probability."

4. Conclusions and future developments

The main results of this chapter, accessible to a second-year undergraduate, are Corollaries 2.3 and 2.5. They follow from the Borel-Cantelli lemmas and Boole's inequality, respectively. Corollary 2.3 states that in an infinite sequence of keystrokes, any fixed-length "work" appears infinitely often with probability 1. Most undergraduates that the author has taught have great difficulty believing this fact, since most statistical tests, for example, are based on finite samples. Corollary 2.5 goes one step further, proving that *every* finite-length piece of work, even those yet unwritten, will *each* appear infinitely often with probability 1. The undergraduate reader will undoubtedly appreciate the "power of infinity" on reading this chapter, while graduate students will enjoy a nonpractical yet deep application of the Borel-Cantelli lemmas.

Example 4 makes a contrast between the finite situation and the infinite one. An important practical problem in this regard would be to use Poisson approximations as in [10] to find the approximate probability that a specific work occurs x times in n keystrokes and to use this process as the basis of a statistical test for randomness.

Author details

Anant P. Godbole
East Tennessee State University, Johnson City, USA

*Address all correspondence to: godbolea@etsu.edu

References

[1] Gamow G. One Two Three Infinity. New York: Bantam Books; 1971

[2] Ash RB. Real Analysis and Probability. New York: Academic Press; 1972

[3] Borel E. Les probabilités dénombrables et leurs applications arithmétiques. Rendiconti del Circolo Matematico di Palermo. 1909;**27**:247-271

[4] Cantelli FP. Su due applicazioni di un teorema di G. Boole. Reale Academia Nazionale dei Lincei. 1917;**26**:295-302

[5] Balakrishnan N, Stepanov A. Generalization of the Borel-Cantelli lemma. The Mathematical Scientist. 2010;**35**:6162

[6] Stepanov A. On strong convergence. Communications in Statistics, Theory and Methods. 2015;**44**:1615-1620

[7] Petrov VV. A generalization of the Borel-Cantelli lemma. Statistics & Probability Letters. 2004;**67**:233-239

[8] Billingsley P. Probability and Measure. New York: John Wiley & Sons, Inc.; 1978

[9] Gnedenko BV, Kolmogorov AN. Limit Theorems for Sums of Independent Random Variables. Boston: Addison Wesley; 1954

[10] Godbole A, Schaffner A. Improved Poisson approximations for word patterns. Advances in Applied Probability. 1993;**25**:334-347

Chapter 4

Prime Numbers Distribution Line

Shcherbakov Aleksandr Gennadiyevich

Abstract

During the analysis of the fractal-primorial periodicity of the natural series of numbers, presented in the form of an alternation (sequence) of prime numbers (1 smallest prime factor > 1 of any integer), the regularity of prime numbers distribution was revealed. That is, the theorem is proved that for any integer = N on the segment of the natural series of numbers from 1 to N + $2\sqrt{N}$: (1) prime numbers are arranged in groups, by exactly three consecutive prime numbers of the form: (P_1-P_2-P_3). In this case, the distance from the first to the third prime number of any group is less than $2\sqrt{N}$ integers, that is, P_3–P_1 < $2\sqrt{N}$ integers. (2) These same prime numbers are redistributed in a line in groups, by exactly two consecutive prime numbers, on all segments of the natural series of numbers shorter than $2\sqrt{N}$ integers.

Keywords: residue groups, prime numbers, primorial, sieve of Eratosthenes, alternations, fractal

1. Introduction

1.1 Line-symmetrical primary-repeatable fractals of the positive integers

In the scientific works [7 pp. 142–147, 8 pp. 77–84, 9 pp. 109–116], positive integers are analyzed, hereinafter the P.I. is represented only as the alternance (array) of primes (according to the 1st least prime factor > 1 from every whole number). Type: 12.3.5.7.3.11.13.3.17.19.3.23.5.3.29 ... 3.*p*.*p*.3.*p* ... 3.*p*.*p*.3.*p* .... with for every recurrent prime = P_1, sieve of Eratosthenes formats the P.I., represented by alternance (array) of the first primes ≤P_1, in the form line-symmetrical repeating fractal-like structure, situated in the section of P.I. from 1 to $P_1\#$, with "eliminated" sections of P.I. and $\varphi(P_1\#)$ not eliminated odd numbers are line-symmetrical to the number = $P_1\#/2$ and are repeated without rearrangement of their position with the period = $P_1\#$, on the basis of rhythmical repeating of two even numbers. Every recurrent prime has its own line-symmetrical primary-repeatable fractal = P_1, then goes fractal = $P_1\#$ (see line 1 **Table 1**).

Every recurrent line-symmetrical fractal -$P_1\#$ is situated on the section of P.I. from 1 to $P_1\#$ and contains $\varphi(P_1\#)$ of the not eliminated odd numbers that are $\varphi(P_1\#)$ of the least residue, belonging to the indicated residue system (I.R.S) according to mod ($P_1\#$), type: C_n to the left from the number = $P_1\#/2$ and ($P_1\#$–C_n) to the right from the number = $P_1\#/2$, with C_n – is residue according to mod($P_1\#$). Hereinafter with the term mod($P_1\#$), we shall indicate the period of fractal $P_1\#$ repetition (I.R.S, sieve of Eratosthenes), equal to product of all first primes ≤P_1 (primorial = $P_1\#$) [1–6].

$C_1 = 1$	3,5,7..	$C_2 = P_2$	**ppp**	..C_3 ..	**ppp**	.. **Cn** ..	**ppp**	$P_1\#$–**Cn**	**ppp**	$P_1\#$–C_2	..7,5.3	($P_1\#$–1)
1 + $P_1\#$	3,5,7..	$C_2 + P_1\#$	ppp	$C_3 + P_1\#$	ppp	Cn + $P_1\#$	ppp	$2P_1\#$–Cn	ppp	$2P_1\#$–C_2	..7,5.3	**($2P_1\#$–1)**
1 + $2P_1\#$	3,5,7..	$C_2 + 2P_1\#$	ppp	$C_3 + 2P_1\#$	ppp	Cn + $2P_1\#$	ppp	$3P_1\#$–Cn	ppp	$3P_1\#$–C_2	..7,5.3	**($3P_1\#$–1)**
...	3,5,7..	...	ppp	...	ppp	...	ppp	...	ppp	...	..7,5.3	...
...	3,5,7..	...	ppp	...	ppp	...	ppp	$P_2\#$–Cn	ppp	$P_2\#$–C_2	..7,5.3	**($P_2\#$–1)**
1 + $P_2\#$	3,5,7..	$C_2 + P_2\#$	ppp	$C_3 + P_2\#$	ppp	Cn + $P_2\#$	ppp		ppp		..7,5.3	

And so on, repeating of fractal = $P_1\#$ with the period = $P_1\#$, with: ppp is alternance of $\leq P_1$

Table 1.
P_2 *repeating of periodical fractal* = $\mathbf{P_1\#}$, *including I.R.S. according to the mod*($P_1\#$).

By term residue according to mod (P_1#), we shall indicate every number, NOT eliminated by Sieve of Eratosthenes, not aliquot to the first primes $\leq P_1$.

Alternance $\leq P_1$ is the section of P.I. in the form of array of primes – NOT residues of mod(P_1#), (for the 1 least common factor > 1 from every NOT residue).

Eliminating (according to diagonals) 1 number multiple to P_2 in every column = Cn, we'll get in P_2 lines of **Table 1**: $\varphi(P_1\#)*(P_2 \text{ lines}) - \varphi(P_1\#)_{\text{multiple}}$ $P_2 = \varphi(P_2\#)$ residue of mod(P_2#). Representing the section of P.I. from 1 to P_2# as one line, we'll get the fractal = P_2# with the period of repeating =P_2#. And so on: every recurrent prime = P_n has got its periodical fractal = P_n# with n is the whole. The numerical illustration is indicated in the scientific works [7–10].

Fractal (P_1#)-I.R.S. mod(P_1#) = (first line of **Table 1**).

Fractal (P_2#)-I.R.S. mod(P_2#) = P_2 lines in **Table 1**. -$\varphi(P_1\#)$ numbers multiple to P_2.

Fractal (P_3#) I.R.S. mod(P_3#) = P_3 lines in **Table 2**.-$\varphi(P_2\#)$ numbers multiple to P_3.

Fractal (P_4#) I.R.S. mod(P_4#) = (P_4 repeating of fractal P_3#)–$\varphi(P_3\#)$ numbers multiple to P_4.

Fractal (P_5#) I.R.S. mod(P_5#) = (P_5 repeating of fractal P_4#)–$\varphi(P_4\#)$ numbers multiple to P_5.

and so on according to cumulative primes.

1.2 Purpose and role of the overall length of the of alternance (array) of the all first primes ≤Pn

It is quite obvious that $\varphi(P_n\#)$ of the least residues of mod(P_n#) type = C and (P_n#–C), of every recurrent fractal = P_n#, gradate P.I. as $\varphi(P_n\#)$ "eliminated" sections of P.I. with different lengths of the type: *C*~~..3pp3.~~*C*~~.3pp3.~~*C* ~~3pp3~~ *C*., with ~~..3pp3..~~ "eliminated" sections of P.I. represented as array of "eliminated" NOT residues of mod(P_n#), or un the form of alternance (array) of the first primes $\leq P_n$, (according to the 1st least prime factor >1 from every NOT residue of mod(P_n#)), hereinafter the alternance $\leq P_n$. C – residue of mod(P_n#) (according to the 1st least >P_n from every residue of mod(P_n#)), location from 1 to P_n# is line symmetrical relating to number = P_n#/2. And further, repeated without rearrangement of their position with the period = P_n#. Then, after we define the overall - maximal length of alternance that we can form using the fist primes p $\leq P_n$, (NOT residues of mod (P_n#)), type C_1 ... 3pp3pp3pp3 ... C_2 that is maximal amount of consequent odd numbers = maximal length of alternance -p $\leq P_n$ (one least NOT residue of mod (P_n#) >1 from the number), we can evaluate the distance between every two consequent residues of mod (P_n#) that is between two primes $< (P_{n+1})^2$, according to formula: $(C_2 - C_1) - 2/2$ of the odd numbers ≤ maximal length of the alternance, (maximal amount of NOT residue of mod(P_n#)).

In the scientific works [7 pp. 142–147, 8 pp. 77–84, 9 pp.109–116], the distribution of groups of 4 consequent residues in the form of "pairs of residues every two residue" is analyzed. But we have no information on distribution of groups of 4, 3, and 2 consequent residues of mod(P_n#) for every fractal P_n#.

In this scientific work, the $\varphi(P_n\#)$ of the least residue of mod(P_n#) of every recurrent fractal -P_n# is indexed as continuous sequence of groups: (a) No 4 has got 4 residues, or (b) No 3 has got 3 residues or (c) No 2 has got 2 consequent residues mod(P_n#). These groups No 4-3-2 are analyzed as subgroups with No 4-3-2 consequent residues of mod(P_n#) that are surrounded by the maximal permissible amount of consequent NOT residues of mod(P_n#).

We used the mathematical induction method to define the overall – maximal length of every kind of subgroups No 4, No 3, No 2 and overall maximal length of P.

$C_1 = 1$	3,5,7,,,	$C_2 = P_3$	**ppp**	..C_3..	**ppp**	.. **Cn** ..	**ppp**	**$P_2\#$–Cn**	**ppp**	**$P_2\#$–C_2**	..7,5.3	($P_2\#$–1)
1 + $P_2\#$	**3,5,7,,,**	$C_2 + P_2\#$	ppp	$C_3 + P_2\#$	ppp	Cn + $P_2\#$	ppp	$2P_2\#$–Cn	ppp	$2P_2\#$–C_2	**..7,5.3**	**($2P_2\#$–1)**
1 + $2P_2\#$	**3,5,7,,,**	$C_2 + 2P_2\#$	ppp	$C_3 + 2P_2\#$	ppp	Cn + $2P_2\#$	ppp	$3P_2\#$–Cn	ppp	$3P_2\#$–C_2	**..7,5.3**	**($3P_2\#$–1)**
...	**3,5,7,,,**	...	ppp	...	ppp	...	ppp	...	ppp	...	**..7,5.3**	...
...	**3,5,7,,,**	...	ppp	...	ppp	...	ppp	$P_3\#$–Cn	ppp	$P_3\#$–C_2	**..7,5.3**	**($P_3\#$–1)**
1 + $P_3\#$	**3,5,7,,,**	$C_2 + P_3\#$	ppp	$C_3 + P_3\#$	ppp	Cn + $P_3\#$	ppp	...	ppp	...	**..7,5.3**	...

And so on, repeating of fractal = $P_2\#$ with the period = $P_2\#$, with: ppp is alternance of $\leq P_2$.
Representing the section of P.I. as line from 1 to $P_3\#$ we'll get the fractal $P_3\#$ and so on.

Table 2.
P_3 *repeating of periodical fractal* = **$P_2\#$**, *including I.R.S. according to the mod*($P_2\#$).

I. sections in the form of maximal long alternances of all first primes $\leq P_n$, (that is maximal permissible amount of all NOT residues of mod(P_n#)), situated between two residues from C_A to C_B, between which, as subgroups are situated the groups of residues of mod (P_n#). Type:

a. -No 4: $C_{A..3pp3..}P_{1..}P_{2..}P_{3..}P_{4..3pp3..}C_B$.

b. No 3: $C_{A..3pp3..}P_{1..}P_{2..}P_{3..3pp3..}C_B$.

c. No 2: $C_{A..3\ pp3..}P_{1..}P_{2..3pp3..}C_B$.

As a result, we detected the loopback of these groups rearrangement from No 4 to No 3 up to No 2 according to the growing amount of the modulus, and the primes order distribution is defined.

2. Three groups of "eliminated" sections of every next fractal

It is quite obvious and requires no proof that indexing φ(P_1#) of the least residues of mod(P_1#) of every recurrent fractal-P_1#, or I.R.S mod(P_1#), is by groups, containing strictly 4 elements; three; two consequent residues of mod(P_1#), we have, that every recurrent fractal-P_1# would be represented as array of three groups of the residues of mod(P_1#), between are situated the alternances $\leq P_1$ (with different lengths) – the consequent NOT residues of mod(P_1#), **of types (a), (b), (c)** repeated without changes with period = P_1#.

a. φ(P_1#) groups No 4 containing strictly FOUR consequent residues of mod (P_1#)$_{A,B,C,D}$ **C**, between which the alternances of different amounts of different first primes $\leq P_1$, NOT residues of mod(P_1#), type: $_AC$..3pp3..$_BC$..3pp3..$_CC$ .. 3pp3..$_DC$. with: ..3pp3.. are alternances of the first primes $\leq P_1$ according to the 1st least common factor > 1 from every NOT residue of mod(P_1#). ($_{1-4}$C) – Four consequent residue of mod(P_1#), including the consequent primes of P.I. section from P_1 to $(P_2)^2$ of $_{A,B,C,D}$***P*** type. Further, the fractal = P_1# represented as φ(P_1#) of No 4 groups (4 residues) of mod (P_1#). Three adjoined groups No 4 for every residue = C (**Table 3**).

The length of group No 4, which means amount odd numbers, restricted by every group No 4 from $_AP$ to $_DP$ and from C_1 to C_4, for the mod(P_1#), is $(R_4-2)/2 \leq (P_2-1)$ of the odd numbers with: $R_4 = (_DP - _AP)$, $R_4 = (_4C - _1C)$, $R_4 \leq 2P_2$ (including 1 group of $R_4 = 2P_2$, detailed information is indicated in Section 5) (**Table 4**).

b. φ(P_1#) groups No 3 containing strictly THREE consequent residues of mod (P_1#). $_{A,B,C}$**C**, between which the alternances of different amounts of

1,3,5,7 ...	$_AP$ ..3pp3..$_BP$..3pp3..$_CP$..3pp3..$_DP$. 3pp3...	P_2^2	$_1$C.3pp3.$_4$C.3pp3..	P_1#
3,5,7 ...	———$_BP$..3pp3..$_CP$..3pp3..$_DP$..3pp3..$_AP$ ...	P_2^2	—$_2$C..3pp3..$_5$C..$_2$C	P_1#
5,7 ...	- ———$_CP$..3pp3..$_DP$.3pp3..$_AP$..3pp3..$_BP$	P_2^2	———$_3$C..3pp3..$_6$C..$_3$C	P_1#

And so on, repeating of fractal = P_1# with the period = P_1#, ppp - is alternance of $\leq P_1$.

Table 3.
Fractal = P_1#, represented as φ(P_1#) of No 4 groups (containing 4 residues) of mod(P_1#). Three adjoined groups No 4 for every residue = C.

Type-(a)	**$C_1 = 1$**	7	11	**$C_4 = 13$**	17	19	**$C_7 = 23$**	29	31	$C = 37$	
Type-(b)		$C_2 = 7$	11	13	**$C_5 = 17$**	19	23	**$C_8 = 29$**	31	37	$C = 41$
Type-(c)			**$C_3 = 11$**	13	17	**$C_6 = 19$**	23	29	$C = 31$		
And so on, repeating of fractal =5# with the period = mod(5#)											

Table 4.
The numerical illustration of the fractal =5# in the form of φ(5#) = 8 groups No 4 (containing four residues). The three adjoined groups No 4 for every residue = C with $R_4 \leq 2P_2 = 2^*7$ (consult Section 5).

different first prime $\leq P_1$, NOT residues of mod (P_1#), type: ${}_AC$..3pp3..${}_BC$..3pp3..${}_CC$. with: ..3pp3.. are alternances of the first primes $\leq P_1$ according to the 1st least common factor > 1 from every NOT residue of mod (P_1#). (${}_{1\text{-}3}$C) – three consequent residues of mod(P_1#), including the consequent primes of P.I. section from P_1 to $(P_2)^2$ of ${}_{A,B,C}$***P*** type. Further, the fractal = P_1# represented as φ(P_1#) of No 3 groups (3 residues) of mod(P_1#). Two adjoined groups No 3 for every residue = C (**Table 5**).

With the unknown to us length of the group No 3 from ${}_AP$ to ${}_CP$ and from C_1 to C_3, for the mod (P_1#), is (R_3–2)/2 of the odd numbers with: R_3 = (${}_CP$–${}_AP$)., R_3 = (${}_3$C–${}_1$C)., R_3 =? (it is quite obvious that for mod (P_1#) R_3 < <R_4) (**Table 6**).

c. φ(P_1#) groups No 2 containing strictly TWO consequent residues of mod (P_1#)${}_{A,B}$**C,** between which the alternances of different amounts of different first prime $\leq P_1$, NOT residues of mod(P_1#), type: ${}_AC$..3pp3..${}_BC$. with: ..3pp3.. are alternances of the first primes $\leq P_1$ according to the 1st least common factor > 1 from every NOT residue of mod(P_1#). (${}_{1\text{–}2}$C) – two consequent residue of mod(P_1#), including the consequent primes of P.I. section from P_1 to $(P_2)^2$ of ${}_{A,B}$***P*** type. Further, the fractal = P_1# represented as φ(P_1#) of No 2 groups (2 residues) of mod(P_1#) (**Tables 7–9**).

1.,3.,5.,7..	**${}_AP$..3pp3..${}_BP$ ..3pp3..${}_CP$**	**P_2^2**	**${}_1$C..3pp3..${}_3$C ... 3pp3.. ${}_1$C ... ${}_3$C**	**P_1#**
3.,5.,7 ...	————————.${}_BP$ ..3pp3..${}_CP$..3pp3..${}_AP$..3pp3..	P_2^2	————————${}_2$C..3pp3..${}_4$C.. ${}_2$C .. ${}_4$C	P_1#
And so on, repeating of fractal = P_1# with the period = P_1#, with: ppp is alternance of ≤ P_1.				

Table 5.
Fractal = P_1#, represented as φ(P_1#) of No 3 groups (containing 3 residues) of mod(P_1#). Two adjoined groups No 3 for every residue = C.

Type-(a)	**$C_1 = 1$**	7	$C_3 = 11$	13	**$C_5 = 17$**	19	**$C_7 = 23$**	29	$C = 31$	
Type-(b)		$C_2 = 7$	11	**$C_4 = 13$**	17	**$C_6 = 19$**	23	**$C_8 = 29$**	31	$C = 37$
And so on, repeating of fractal =5# with the period =5#.										

Table 6.
*The numerical illustration of the fractal =5# in the form of φ(5#) = 8 groups No 3 (containing two residues). The two adjoined groups No 3 for every residue = C with R_3 =? (=2*5 consult Section 6).*

1.,3.,5.,7..	**${}_AP$..3pp3..${}_BP$..3pp3..${}_AP$.3pp3..**	**P_2^2**	**${}_1$C..3pp3..${}_2$C..3pp3..${}_1$C..3pp3..${}_2$C..${}_1$C..${}_2$C...**	**P_1#**
And so on, repeating of fractal = P_1# with the period = P_1#, ppp - is alternance of≤ P_1.				

Table 7.
Fractal = P_1#, represented as φ(P_1#) of No 2 groups of mod(P_1#).

1.,3.,5.,7..	${}_AP$..3pp3..${}_BP$..3pp3..${}_AP$.3pp3..	P_3^2	${}_1$C..3pp3..${}_2$C..3pp3..${}_1$C..3pp3..${}_2$C..${}_1$C..${}_2$C...	$P_2\#$
And so on, repeating of fractal = $P_2\#$ with the period = $P_2\#$, ppp - is alternance of $\leq P_2$.				

Table 8.
Fractal = $P_2\#$, represented as $\varphi(P_2\#)$ of No 2 groups of mod$(P_2\#)$.

1.,3.,5.,7..	${}_AP$..3pp3..${}_BP$..3pp3..${}_AP$.3pp3..	P_4^2	${}_1$C..3pp3..${}_2$C..3pp3..${}_1$C..3pp3..${}_2$C..${}_1$C..${}_2$C...	$P_3\#$
And so on, repeating of fractal = $P_3\#$ with the period = $P_3\#$, ppp - is alternance of $\leq P_3$.				

Table 9.
Fractal = $P_3\#$, represented as $\varphi(P_3\#)$ of No 2 groups of mod$(P_3\#)$.

With the unknown to us, length of the group No 2 from ${}_AP$ to ${}_BP$ and from C_1 to C_2, for the mod$(P_1\#)$, is $(R_2–2)/2$ of the odd numbers with: $R_2 = ({}_BP–{}_AP)$., $R_2 = ({}_2C–{}_1C)$, R_2 =? (it is quite obvious that for mod$(P_1\#)$ $R_2 << R_3$).

Herewith for each group No 4-3-2 according to mod$(P_1\#)$, there are two residues of mod$(P_1\#)$: C_A – to the left and C_B – to the right, that is every group No 4-3-2 is the subgroup on the P.I. sections of the length unknown to us from C_A to C_B: (a)C_A-(C_1-C_2-C_3-C_4)-C_B. (b) C_A-(C_1-C_2-C_3)-C_B. (c) C_A-(C_1-C_2)-C_B.

3. Correlations of length limits of the subgroups No 4, No 3, No 2

In the scientific works [7 pp. 142–147, 8 pp. 77–84, 9 p. 109–116, 10 p. 1805], including Section 5 of this work, the overall – maximal length of the subgroup No 4 (containing 4 residual for every recurrent fractal -$P_n\#$), of type is defined:

max $R_4 = (C_4 – C_1) = 2P_{n+1}$ of whole numbers.

Herewith, it is quite obvious and it is beyond argument that relations of limits, unknown to us of groups No 4-3-2 length according to the increasing modulus are indicated in **Table 10**.

Prime value P_n	Fractal - $P_n\#$	Period of fractal repetition =mod ($P_n\#$).	Max length of φ($P_n\#$) of the subgroups No 4 maxR_4 = (C_4–C_1)	>>	Max length φ($P_n\#$) of the subgroups No 3 maxR_3 = (C_3–C_1)	>>	Max length φ($P_n\#$) of the subgroups No 2 maxR_2 = (C_2–C_1)
P_1	$P_1\#$	mod$(P_1\#)$.	max R_4 = (C_4–C_1) = $2P_2$	>>	max R_3 = (C_3–C_1) =?	>>	max R_2 = (C_2–C_1) =?
P_2	$P_2\#$	mod$(P_2\#)$.	max R_4 = (C_4–C_1) = $2P_3$	>>	max R_3 = (C_3–C_1) =?	>>	max R_2 = (C_2–C_1) =?
P_4	$P_4\#$	mod$(P_4\#)$.	max R_4 = (C_4–C_1) = $2P_5$	>>	max R_3 = (C_3–C_1) =?	>>	max R_2 = (C_2–C_1) =?
P_5	$P_5\#$	mod$(P_5\#)$.	max R_4 = (C_4–C_1) = $2P_6$	>>	max R_3 = (C_3–C_1) =?	>>	max R_2 = (C_2–C_1) =?
...	...	...	...	>>	...	>>	...
P_n	$P_n\#$	mod$(P_n\#)$.	max R_4 = (C_4–C_1) = $2P_{n+1}$	>>	max R_3 = (C_3–C_1) =?	>>	max R_2= (C_2–C_1) =?
And so on, repeating of fractal = $P_n\#$ with the period = $P_n\#$. $C_{1–4}$ - residue of mod ($P_n\#$).							

Table 10.
The relation of length limits of the subgroups according to the increasing modulus.

4. Distribution of prime number

Correlation of length limits of the subgroups in **Table 10** and distribution of groups of the indexed residues No 4-3-2, in every respective fractal $-P_n\#$ according to the increasing modulus is defined by theorem 1.

Theorem 1. The loopback of prime number distribution.

Every prescribed prime number squared = $(P_2)^2$ defines the distribution of all previous prime numbers < $(P_2)^2$, as all first prime numbers are less than every prescribed prime number squared = $(P_2)^2$, are situated in the P.I. as part of fractal $P_1\#$, where they are distributed by subgroups of **(a)**, **(b)**, **(c)** types.

a. $\varphi(P_1\#)$ of subgroups No 4 having pure FOUR consequent prime number ($_{A,B,C,D}C$) of $P_1 < ({}_AC\text{-}{}_BC\text{-}{}_CC\text{-}{}_DC) < P_2^2; P_3^2$ type. At the P.I. section from P_1 to P_2^2 (including from $(P_2\text{–}2)^2$ to P_2^2), and further, from P_2^2 to $P_1\#$ pure FOUR consequent residues of mod$(P_1\#)$ on all P.I. sections with length not exceeding $2P_3$ of whole numbers with length of every subgroup No 4 at every section is:

$$R_4 = ({}_DC\text{–}{}_AC) \leq 2P_2.$$

In that case, these primes of fractal = $\boldsymbol{P_1\#}$, by loopback, are distributed by groups:

b. $\varphi(P_1\#)$ of subgroups No 3 having pure THREE consequent prime number ($_{A,B,C}C$) of $P_1 < ({}_AC\text{-}{}_BC\text{-}{}_CC) < P_2^2$ type. At the P.I. section from P_1 to P_2^2 and further, from P_2^2 to $P_1\#$ pure THREE consequent residues of mod$(P_1\#)$ on all P.I. sections with length not exceeding $2P_2$ of whole numbers with length of every subgroup No 3 at every section is: $R_3 = ({}_CC\text{–}{}_AC) \leq \boldsymbol{2P_1}$.

c. $\varphi(P_1\#)$ of subgroups No 2 having pure TWO consequent prime number ($_{A,B}C$) of P_0, $P_1 < ({}_AC\text{-}{}_BC) < P_2^2$ type. At the P.I. section from P_1 to P_2^2 and further, from P_2^2 to $P_1\#$ pure TWO consequent residues of mod$(P_1\#)$ on all P. I. sections with length not exceeding $2P_1$ of whole numbers with length of every subgroup No 2 at every section is: $R_2 = ({}_BC\text{–}{}_AC) \leq \boldsymbol{2P_0}$.

With: $_{A,B,C,D}C$ are consequent residues of mod$(P_1\#)$ including the primes < $(P_2)^2$. $R_{4\text{-}3\text{-}2}$ is the remainder of the first and the last number of every group No 4-3-2 (of the fractal $P_1\#$). Further, the length of the subgroup No 4-3-2 as the amount of odd numbers, restricted by every group from $_AC$ to $_{B\text{-}C\text{-}D}C$, from $_AP$ to $_{B\text{-}C\text{-}D}P$ are $(R_{4,3,2}\text{–}2)/2$ odd numbers.

The order of groups (a), (b), (c) rearrangement according to the increasing modulus for visual clarity is indicated in **Table 11.**

5. Proof of theorem

5.1 Proof of section a of the Theorem 1

It is feasible that in P.I. using the first prime number $\leq P_n$ (NOT residues of mod $(P_n\#)$), by the only single way, we can form the maximal long P.I. section as the maximal long alternance – array for the 1 least common factor > 1 from every NOT residue of mod$(P_n\#)$. That is that maximal amount of NOT residues of mod$(P_n\#)$, maximal long alternance $\leq \boldsymbol{P_n}$.

1	2		3		4		5	
Fractal from 1 to $P_n\#$ Composition and its repeating =mod($P_n\#$)	Length of P.I. section which defines values of primes number of this fractal including on the P.I section: $(P_{n+1})^2–(P_{n+1}–2)^2 = 4P_{n+1}$ of whole numbers		φ($P_n\#$) groups No 4 for 4 residues of mod ($P_n\#$) (containing 4 simple) $_AC$-$_BC$-$_CC$-$_DC$ $< P_{n+1}^2$		φ($P_n\#$) groups No 3 for 3 residues of mod ($P_n\#$) (containing 3 simple) $_AC$-$_BC$-$_CC$ $< P_{n+1}^2$		φ($P_n\#$) groups No 2 for 2 residues of mod ($P_n\#$) (containing 2 simple) ($_AC$-$_BC$) $< P_{n+1}^2$	
			Subgroup No 4 length. $R_4 = {}_DC–{}_AC$	Length of every section for the group No 4	Subgroup No 3 length. $R_3 = {}_CC–{}_AC$	Length of every section for the group No 3	Subgroup No 2 length. $R_2 = {}_BC–{}_AC$	Length of every section for the group No 2
1 ÷ $P_n\#$ mod($P_n\#$)	from $(P_n)^2$ to $(P_{n+1})^2$	$\geq 4P_{n+1}$ number	$R_4 \leq 2P_{n+1} < (P_n\# \pm P_{n+1})$	$\leq 2P_{n+2} <$ $(P_n\# \pm P_{n+2})$	$R_3 \leq 2P_n$ **(Table 12)**	$\leq 2P_{n+1}$ number	$R_2 \leq 2P_{n-1}$ **(Table 13)**	$\leq 2P_n$ number
…	…	…	…	…	…	…	…	…
1 ÷ $P_1\#$ mod($P_1\#$)	from $(P_1)^2$ to $(P_2)^2$	$\geq 4P_2$ number	$R_4 \leq 2P_2$	$\leq 2P_3$ number	$R_3 \leq 2P_1$	$\leq 2P_2$ number	$R_2 \leq 2P_0$	$\leq 2P_1$ number
1 ÷ $P_2\#$ mod($P_2\#$)	from $(P_2)^2$ to $(P_3)^2$	$\geq 4P_3$ number	$R_4 \leq 2P_3$	$\leq 2P_4$ number	$R_3 \leq 2P_2$	$\leq 2P_3$ number	$R_2 \leq 2P_1$	$\leq 2P_2$ number
1 ÷ $P_3\#$ mod($P_3\#$)	from $(P_3)^2$ to $(P_4)^2$	$\geq 4P_4$ number	$R_4 \leq 2P_4$	$\leq 2P_5$ number	$R_3 \leq 2P_3$	$\leq 2P_4$ number	$R_2 \leq 2P_2$	$\leq 2P_3$ number
1 ÷ $P_4\#$ mod($P_4\#$)	from $(P_4)^2$ to $(P_5)^2$	$\geq 4P_5$ number	$R_4 \leq 2P_5$	$\leq 2P_6$ number	$R_3 \leq 2P_4$	$\leq 2P_5$ number	$R_2 \leq 2P_3$	$\leq 2P_4$ number
1 ÷ $P_5\#$ mod($P_5\#$)	from $(P_5)^2$ to $(P_6)^2$	$\geq 4P_6$ number	$R_4 \leq 2P_6$	$\leq 2P_7$ number	$R_3 \leq 2P_5$	$\leq 2P_6$ number	$R_2 \leq 2P_4$	$\leq 2P_5$ number
… and so on according to the increasing meanings of the modulus …								

Table 11.
Loopback of primes number subgroups distribution according to the increasing meanings of the modulus.

Then for every recurrent prime number $P_n = P_1$ at the P.I., formed as recurrent line symmetrical, primary-repeatable periodical fractal = P_1# or I.R.S. according to mod(P_1#), (see the first line of **Table 1**), the maximal long P.I. section, formed as the alternance of the all first primes number $\leq P_1$, (NOT residues of mod(P_n#)), shall be situated within the P.I. section from C_A to C_B, with as the subgroup is the only maximal long maximal subgroup No 4 (C_1-C_2-C_3-C_4) with the 4 consequent residue of mod(P_1#): Type: $C_A = (P_1\#–P_3)..C_1 = (P_1\#–P_2)$.

... $C_2 = (P_1\#–1), C_3 = (P_1\# + 1)$ ... $C_4 = (P_1\# + P_2)$ ... $C_B = (P_1\# + P_3)$. The length of such maximally long subgroup No 4 of mod(P_1#), is: max $R_4 = (C_4–C_1) = (P_1\# + P_2)–(P_1\#–P_2) = 2P_2$ of the whole numbers.

The limit of length of the P.I. section within which from C_A to C_B would be situated maximal as well as all the other φ(P_1#) subgroups No 4 (with 4 residues) of mod(P_1#), is: $(C_B–C_A) = (P_1\# + P_3)–(P_1\#–P_3) = 2P_3$ of the whole numbers.

It is genuinely:

At the line-symmetrical, primary-repeatable fractal-P_1# or I.R.S. according to mod(P_1#), φ(P_1#) of the least residue (indexed in the form of φ(P_1#) groups No 4 of mod(P_1#), with the alternances $\leq P_1$ with different lengths), are situated line-symmetrically relating to the center of symmetry of the fractal-P_1#, of the number = (P_1#/2). That is they are situated reflecting in pairs and are formed by two different ways: (the left and the right sieve of Eratosthenes), to the left and to the right from the symmetry center of the fractal = P_1# of number = P_1#/2. To the right – for the increasing values numbers of the P.I. from P_1#/2 to P_1# to the left for the decreasing values of the P.I. P_1#/2 up to 1.

To every left group No 4 of mod(P_1#), with the remainder $R_4 = (C_4–C_1)$, is matched by line-symmetrical right group No 4 mod(P_1#), with reflecting location of the same first primes number in the same amount and of the same length of the alternance $\leq P_1$: $R_4 = (P_1\#–C_1)–(P_1\#–C_4) = (C_4–C_1)$ consult [7 pp. 142–147, 8 pp. 77–84, 9 pp. 109–116].

Besides two not reflecting that is formed solely subgroup of group No 4: with constant reminder for every P_n of type: $R_4 = (P_1\#/2 + 4)–(P_1\#/2–4) = 8$.

And the section of P.I. fractal P_1# (I.R.S. of mod(P_1#) from C_A to C_B), represented by alternance $\leq P_1$ with using of all NOT residues of mod(P_1#), (according to the 1 the least > 1), with the subgroup is situated the only maximally long - maximal group No 4 with 4 residues of mod(P_1#) (C_1-C_2-C_3-C_4). Type: $C_A = (P_1\#\text{-}P_3)$ ... 3pp3 ... $C_1 = (P_1\#\text{-}P_2)$ ... 3pp3 ... $C_2 = (P_1\#\text{-}1)$, $C_3 = (P_1\# + 1)$ ... 3pp3 ... $C_4 = (P_1\# + P_2)$ ... ... 3pp3 ... $C_B = (P_1\# + P_3)$.

Thus in the fractal-P_1#-I.R.S. of the mod(P_1#), there is only one maximally long subgroup No 4, situated within the maximally long alternance $\leq P_1$, using all NOT residues of mod(P_1#), at the P.I. section ($P_1\# \pm P_2$) with length maximal $R_4 = 2P_2$ restricting $(R_4–2)/2 = (2P_2–2)/2 = (P_2–1)$ of the odd numbers, situated within the P. I. section, formed solely from $(P_1\#–P_3)$ to $(P_1\# + P_3)$ with length of $(P_3–1)$ of odd numbers.

It is quite obvious that all the other, line-symmetrical subgroups No 4 of mod (P_1#), situated within the alternances $\leq P_1$ with different lengths or NOT residues, of mod(P_1#), cannot have the maximal length as they are formed by two different ways, that is they would be shorter than $R_4 < 2P_2$, and situated within the P.I. sections from C_A to C_B with length not exceeding the maximal long P.I. section $(C_B–C_A) \leq 2P_3$ of the whole numbers, not exceeding $(P_3–1)$ of the odd numbers.

And so on, for all posterior prime numbers = P_n, at the increasing fractals -P_n# with n - as the whole number and proves the reality of the values of column No 3 of **Table 11** and item (a) of Theorem 1.

It is feasible that there is such a prime number $P_n = P_{(1)}$, for which the P.I. is the line-symmetrical fractal $P_{(1)}$#, situated at the P.I. section from 1 to $P_{(1)}$# with subgroup No 4 (containing 4 residues) of mod($P_{(1)}\#$) with length = $R_4 > 2^*P_{(2)}$, situated within the alternance of all first primes number $\leq P_{(1)}$ within the P.I. section with length $> 2^*P_{(3)}$, $>(P_{(3)}-1)$ of the odd numbers. Then every subgroup No 4 would be line-symmetrical to the left and to the right from the center of the fractal $P_{(1)}\#$ symmetry of the number $P_{(1)}\#/2$. That is, in the result, we'll get in fractal $P_{(1)}\#$ using all primes number $\leq P_{(1)}$, – we can by more than by one way from the maximally long alternance of all the prime numbers $\leq P_{(1)}$, that is by the sieve of Eratosthenes, focused to the left and to the right (to the left and to the right from the number = $P_{(1)}\#/2$), that is contrary to the taken axiom.

6. The maximal length of P.I. section with maximal long subgroups No 3 (with 3 residues) for the mod($P_2\#$)

At the fractal -$P_1\#$, there are $\varphi(P_1\#)$ subgroups No 4 of mod($P_1\#$), with length $R_4 \leq 2P_2$ of the whole numbers including one maximal long subgroup No 4 of mod ($P_1\#$) with length max $R_4 = 2P_2$ of the whole numbers. At the transition from mod ($P_1\#$) to mod($P_2\#$), the fractal $P_1\#$ and $\varphi(P_1\#)$ of groups No 4 repeat P_2 times. Then at the P.I. section from 1 to $P_2\#$ (at P_2 lines of **Table 1**), we'll get $\varphi(P_1\#)$ columns of groups No 4 of mod($P_1\#$), with length $R_4 \leq 2P_2$, (P_2 lines at the column No 4).

It is quite obvious that in Section 8.1, it is proved that if by number P_2, "eliminate," that is moved to mod($P_2\#$) 1 time every elimination C_2 and C_3, in the column of every $\varphi(P_1\#)$ group No 4 of mod($P_1\#$), than at the P.I. section from 1 to $P_2\#$, (that is at the fractal $P_2\#$), we'll get $=2\varphi(P_1\#)$ groups No 3 of mod($P_2\#$) of the same length, that is $R_4 \leq 2P_2$ of mod($P_1\#$) would become = $R_3 \leq 2P_2$ of mod($P_2\#$) with changing of the alternances composition from $\leq P_1$ to $\leq P_2$.

As all one by one eliminated residues C_2 or C_3, at rearrangement of the groups from No 4 to No3 for the mod($P_2\#$), cannot change the length of none of the subgroups, that is all R_3 would permanently be $\leq 2P_2$, included in $2\varphi(P_1\#)$ groups No 3 of mod ($P_2\#$) there is only one P_2 times repeated, maximally long subgroup No 4 of mod($P_1\#$) with the alternance $\leq P_1$, with length max $R_4 = 2P_2$, that would be

(a) Fractal = $P_2\#$, $\varphi(P_2\#)$ group No 3 of mod($P_2\#$), alternance $\leq P_2$, maxR_3 = **2**P_2, with n = (multiple $P_2 \pm 1)/P_1\#$ = the whole. At the P.I. sections ($nP_1\# \pm P_2$) and $(P_2-n)P_1\# \pm P_2$. Within the limits of P.I. section ($nP_1\# \pm P_3$) with: n and (P_2–n) is line number in **Table 1**								
....C_A....	.	......C_1...	11,	$(\text{кр.}P_2)$ и C_2	3,	... ... C_3...	..	....C_B... ...
$=nP_1\#-P_3$	.	$=nP_1\#-P_2$	3,	$=nP_1\#. \pm .1$	5,	$=nP_1\# + P_2$		$=nP_1\# + P_3$
${}_AC = P_2\#-C_B$		${}_1C = P_2\#-C_3$	7,	${}_2C$ и $(\text{кр.}P_2)$	7,	${}_3C = P_2\#-C_1$		${}_BC = P_2\#-C_A$
$(P_2-n)P_1\#-P_3$		$(P_2-n)P_1\#-P_2$	5,	$(P_2-n)P_1\# \pm 1$	3,	$(P_2-n)P_1\# + P_2$		$(P_2-n)P_1\# + P_3$
			3		11.			
(b) Fractal = $P_3\#$, $\varphi(P_3\#)$ group No 3 of mod($P_3\#$), alternance $\leq P_3$, maxR_3 = **2**P_3, with n = (multiple $P_3 \pm 1)/P_2\#$ = the whole. At the P.I. sections ($nP_2\# \pm P_3$) and $(P_3-n)P_2\# \pm P_3$. Within the limits of P.I. section ($nP_2\# \pm P_4$) with: n and (P_3–n) is line number on **Table 2**								
...C_A....	.	......C_1...	11,	$(\text{кр.}P_3)$ и C_2	3,	... ... C_3...	..	....C_B... ...
$=nP_2\#-P_4$	.	$=nP_2\#-P_3$	3,	$=nP_2\#. \pm .1$	5,	$=nP_2\# + P_3$		$=nP_2\# + P_4$
${}_AC = P_3\#-C_B$		${}_1C = P_2\#-C_3$	7,	${}_2C$ и $(\text{кр.}P_3)$	7,	${}_3C = P_2\#-C_1$		${}_BC = P_2\#-C_A$
$(P_3-n)P_2\#-P_4$		$(P_3-n)P_2\#-P_3$	5,	$(P_3-n)P_2\# \pm 1$	3,	$(P_3-n)P_2\# + P_3$		$(P_3-n)P_2\# + P_4$
			3		11,			
(c) And so on for every mod($P_n\#$), max$R_3 = 2P_n$, n = $(\text{кр.}P_n \pm 1)/P_{n-1}\#$ = the whole, $P_{n-(1)}$-primes								

Table 12.
*Type and formula for indexing of two line-symmetrical, maximally long subgroups No 3 (having 3 residues) at the increasing fractal according to the increasing modulus (**Tables 11** and **14**).*

(b) Fractal = $P_3\#$, $\varphi(P_3\#)$ group No 2 of mod($P_3\#$), alternance $\leq P_3$, maxR_2 = $2P_2$, with n = (multiple $P_2{\cdot}P_3 \pm 1)/P_1\#$ = the whole. At the P.I. sections (n$P_1\# \pm P_2$) and ($P_2{\cdot}P_3$–n)$P_1\# \pm P_2$.
Within the limits of P.I. section (n$P_1\# \pm P_3$)., with: n and ($P_2{\cdot}P_3$–n) is line number on **Table 1**

… …C_A ….	.	… …C_1 …	..	(кр.P_2) (кр.P_3)	3,	… … C_2 …	.	… .C_B … …
=n$P_1\#$–P_3	.	=n$P_1\#$–P_2	..	=n$P_1\#$. ± .1	5,	=n$P_1\#$ + P_2	.	=n$P_1\#$ + P_3
$_AC = P_3\#–C_B$		$_1C = P_3\#–C_3$	7,	(кр.P_3) (кр.P_2)	7.	$_2C = P_3\#–C_1$		$_BC = P_3\#–C_A$
(P_2P_3–n)$P_1\#$–		(P_2P_3–n)$P_1\#$–	5,	(P_2P_3–n)$P_1\# \pm 1$	..	(P_2P_3–		(P_2P_3–n)$P_1\#$ + P_3
P_3		P_2	3		..	n)$P_1\#$ + P_2		

(c) Fractal = $P_4\#$, $\varphi(P_4\#)$ group No 2 of mod($P_4\#$), alternance $\leq P_4$, maxR_2 = $2P_3$, with n = (multiple $P_4{\cdot}P_3 \pm 1)/P_2\#$ = the whole. At the P.I. sections (n$P_2\# \pm P_3$) and ($P_4{\cdot}P_3$–n)$P_2\# \pm P_3$.
Within the limits of P.I. section (n$P_2\# \pm P_4$)., with: n-and ($P_4{\cdot}P_3$–n) is line number on **Table 2**

… …C_A ….	.	… …C_1 …	..	(кр.P_4) (кр.P_3)	3,	… … C_2 …	.	… .C_B … …
=n$P_2\#$–P_4	.	=n$P_2\#$–P_3	..	=n$P_2\#$. ± .1	5,	=n$P_2\#$ + P_3	.	=n$P_2\#$ + P_4
$_AC = P_4\#–C_B$		$_1C = P_4\#–C_3$	7,	(кр.P_3) (кр.P_4)	7.	$_2C = P_4\#–C_1$		$_BC = P_4\#–C_A$
(P_4P_3–n)$P_2\#$–		(P_4P_3–n)$P_2\#$–	5,	(P_4P_3–	..	(P_4P_3–		(P_4P_3–
P_4		P_3	3	n)$P_2\# \pm 1$	..	n)$P_2\#$ + P_3		n)$P_2\#$ + P_4

(d) And so on for every mod($P_{n+1}\#$), maxR_2 = $2P_n$, n = $(P_{n+1}{\cdot}P_n \pm 1)/P_{n-1}\#$ = the whole, P_n-primes

Table 13.
*Type and formula for indexing of two line-symmetrical, maximally long subgroups No 2 (having 2 residues) at the increasing fractal according to the increasing modulus (**Tables 11** and **14**).*

restructured into two maximally long line-symmetrical groups No 3 of mod($P_2\#$) by "eliminating" the residues C_2 and C_3 by number multiple to P_2, (1 time at P_2 lines). As all the other $2\varphi(P_1\#)$–2 subgroups, changed from No 4 to No 3 for mod($P_2\#$) are shorter than (P_2–1) of the odd numbers that is: $R_3 < 2P_2$. In Sections 8.1 and 8.3, there are no other ways of making or changing the subgroups No 3 of mod($P_2\#$) with length $R_3 > 2P_2$.

Order, type, and formula of indexing of two subgroups No 3 according to the increasing modulus are represented in **Table 12** a, b, c.

The length of these two line-symmetrical subgroups No 3 of mod($P_2\#$), that is the length of alternance $\leq P_2$ from C_1 to C_3, is maximal R_3 = (C_3–C_1) = (n$P_1\#$. ± P_2)–(n$P_1\#$.–P_2) =$2P_2$; (P_2–1) of the odd numbers. Two of these subgroups No 3 are situated within P.I. section (n$P_1\#$. ± P_3) with length (C_B–C_A) = $2P_3$; (P_3–1) of the odd numbers from C_A to C_B. Numerical values of these two maximal subgroups No 3 are defined according to the formula (multiple P_2 and C_2 = multiple P_2 ± 2) is (n$P_1\#$. ± .1) and (P_2–n)$P_1\#$ ± 1, with n and (P_2–n) define the number of the line for the group No 3 of mod($P_2\#$) in column P_2 and the repeated maximal of the group No 4 of mod($P_1\#$) with the period = $P_1\#$ (consult **Table 1**). That is n = (multiple P_2. ± 1)/$P_1\#$ = the whole < P_2/2.

Whereas, it is quite obvious that in and proved in Section 8.2, the other subgroups No 3 of mod($P_2\#$), with different lengths, changed from groups No 4 would be within P.I. section, with limit length = $2P_3$ of the whole numbers.

And so on, for every of all posterior primes = P_n, at the increasing fractals $P_n\#$, with n is the whole, represented in **Tables 11** and **14** (the proof is indicated in Section 10). (Numerical illustrations are in **Table 15**).

7. The maximal length of the P.I. section, where two maximally long subgroups No 2 (with 2 residues) for the mod($P_3\#$)

Representing as one line, the first P_2 lines in **Table 1** we'll get the fractal $P_2\#$ according to the mod($P_2\#$) – I.R.S. at mod($P_2\#$), that is situated at the P.I. section

1	2	3	4
Fractal composition - $P_n\#$	$\varphi(P_n\#)$ groups No 4 of mod $(P_n\#)$. Including 1 group = maxR_4 **Proof in Section 5**	Loopback (P_n-3) $\varphi(P_{n-1}\#)$ groups No 3 + loopback $2\varphi(P_{n-1}\#)$ group No 4 [mod$P_{n-1}\#$]= = $\varphi(P_n\#)$ groups No 3 of mod$(P_n\#)$ including 2 groups No 3 = max R_3 of mod $(P_n\#)$ **Proved in Sections 6 and 8**	Loopback (P_n-2) $\varphi(P_{n-1}\#)$ Groups No 2 + loopback $\varphi(P_{n-1}\#)$ groups No 3 [mod$P_{n-1}\#$]= =$\varphi(P_n\#)$ groups No 2 of mod$(P_n\#)$ including 2 groups No 2 = max R_2 mod $(P_n\#)$ **Proved in Sections 7 and 9**
By repeating P_1 times the line of fractal -$P_0\#$ and group No 4-3-2, we'll get $\varphi(P_0\#)$ columns of groups No 4-3-2 of mod$(P_0\#)$ within the alternances $\le P_0$ (P_1 line in the column). By "eliminating" 1 number multiple -P_1 (in line of every column No 4-3-2), that is by transiting this group for mod$(P_1\#)$ with changing of its length from $R_{(4,3)}$ to $R_{3,2}$ and alternances composition from $\le P_0$ to $\le P_1$. At the P.I. section from 1 to $P_1\#$ we'll get the fractal -$P_1\#$ of mod$(P_1\#)$. $\varphi(P_1\#)$ groups No 4-3-2			
Amount of groups of Length and location	$\varphi(P_1\#)$ groups No 4. R_4. $\le 2P_2$ at every sectione $<2P_3$ is group No 4	$\varphi(P_0\#)^*$ + $2\varphi(P_0\#)$ = $\varphi(P_1\#)$ $^*(P_1-3)$ $R_{(4)} \le 2P_1$ group $R_{(3)} \le 2P_0$ No 3 $R_3 \le 2P_1$ At every section $<2P_2$ is group No 3	$\varphi(P_0\#)^*$ + $\varphi(P_0\#)$ = $\varphi(P_1\#)$ $^*(P_1-2)$ $R_{(3)} \le 2P_0$ group $R_{(2)} \le 2P_{..}$ No 2 $R_2 \le 2P_0$ At every section $<2P_1$ is group No 2
Formula = n Type max R Groups and location	$(P_1\# \pm P_2)$ max $R_4 = 2P_2$ At the section with length = $(P_1\# \pm P_3)$	n = (multiple $P_1 \pm 1$) / $P_0\#$ = the whole $(nP_0\#. \pm .P_1)$ and $(P_1-n)P_0\#. \pm .P_1$ 2 groups No 3- max $R_3 = 2P_1$. On the segment of length = $(nP_0\# \pm P_2)$	n = (multiple $P_0{}^*P_1 \pm 1)/P_{(.)}\#$ = the whole $(nP_{(.)}\# \pm P_0)$ and $(P_0P_1-n)P_{(.)}\# \pm P_0$ 2 groups No2- max $R_2 = 2P_0$. On the segment of length = $(nP_{(..)}\# \pm P_1)$
By repeating P_2 times the line of fractal –$P_1\#$ and group No 4-3-2, we'll get $\varphi(P_1\#)$ columns of groups No 4-3-2 of mod$(P_1\#)$ within the alternances $\le P_1$ (P_2 line in the column). By "eliminating" 1 number multiple -P_2 (in line of every column No 4-3-2), that is by transiting this group for mod$(P_2\#)$ with changing of its length from $R_{(4,3)}$ to $R_{3,2}$ and alternances composition from $\le P_1$ to $\le P_2$. At the P.I. section from 1 to $P_2\#$ we'll get the fractal –$P_2\#$ of mod$(P_2\#)$. $\varphi(P_2\#)$ groups No 4-3-2			
Amount of groups of length and location	$\varphi(P_2\#)$ groups No 4. R_4. $\le 2P_3$ at every sectione $<$ $2P_4$ is group No 4	$\varphi(P_1\#)^*$ + $2\varphi(P_1\#)$ = $\varphi(P_2\#)$ $^*(P_2-3)$ $R_{(4)} \le 2P_2$ group $R_{(3)} \le 2P_1$ No 3 $R_3 \le 2P_2$ At every section $<2P_3$ is group No 3	$\varphi(P_1\#)^*$ + $\varphi(P_1\#)$ = $\varphi(P_2\#)$ $^*(P_2-2)$ $R_{(3)} \le 2P_1$ group $R_{(2)} \le 2P_0$ No 2 $R_2 \le 2P_1$ At every section $<2P_2$ is group No 2
Formula = n Type max R Groups and location	$(P_2\# \pm P_3)$ max $R_4 = 2P_3$ At the section with length = $(P_2\# \pm P_4)$	n = (multiple $P_2 \pm 1$) / $P_1\#$ = the whole $(nP_1\#. \pm .P_2)$ and $(P_2-n)P_1\#. \pm .P_2$ 2 groups No 3- max $R_3 = 2P_2$. On the segment of length = $(nP_1\# \pm P_3)$	n = (multiple $P_2{}^*P_1 \pm 1)/P_0\#$ = the whole $(nP_0\# \pm P_1)$ and $(P_2P_1-n)P_0\# \pm P_1$ 2 groups No2- max $R_2 = 2P_1$. On the segment of length = $(nP_0\# \pm P_2)$
By repeating P_3 times the line of fractal –$P_2\#$ and group No 4-3-2, we'll get $\varphi(P_2\#)$ columns of groups No 4-3-2 of mod$(P_2\#)$ within the alternances $\le P_2$ (P_3 line in the column). By "eliminating" 1 number multiple–P_3 (in line of every column No 4–3-2), that is by transiting this group for mod$(P_3\#)$ with changing of its length from $R_{(4,3)}$ to $R_{3,2}$ and alternances composition from $\le P_2$ to $\le P_3$. At the P.I. section from 1 to $P_3\#$ we'll get the fractal –$P_3\#$ of mod$(P_3\#)$. $\varphi(P_3\#)$ groups No 4-3-2:			
Amount of groups of length and location	$\varphi(P_3\#)$ groups No 4. R_4. $\le 2P_4$ at every section $<2P_5$ is group No 4	$\varphi(P_2\#)^*$ + $2\varphi(P_2\#)$ = $\varphi(P_3\#)$ $^*(P_3-3)$ $R_{(4)} \le 2P_3$ group $R_{(3)} \le 2P_2$ No 3 $R_3 \le 2P_3$ At every section $<2P_4$ is group No 3	$\varphi(P_2\#)^*$ + $\varphi(P_2\#)$ = $\varphi(P_3\#)$ $^*(P_3-2)$ $R_{(3)} \le 2P_2$ group $R_{(2)} \le 2P_1$ No 2 $R_2 \le 2P_2$ At every section $<2P_3$ is group No 2
Formula = n Type max R Groups and location	$(P_3\# \pm P_4)$ max $R_4 = 2P_4$ At the section with length = $(P_3\# \pm P_5)$	n = (multiple $P_3 \pm 1$) / $P_2\#$ = the whole $(nP_2\#. \pm .P_3)$ and $(P_3-n)P_2\#. \pm .P_3$ 2 groups No 3- max $R_3 = 2P_3$. On the segment of length = $(nP_2\# \pm P_4)$	n = (multiple $P_2{}^*P_3 \pm 1)/P_1\#$ = the whole $(nP_1\# \pm P_2)$ and $(P_2P_3-n)P_1\# \pm P_2$ 2 groups No 2- max $R_2 = 2P_2$. On the segment of length = $(nP_1\# \pm P_3)$
…	…	…	…
And so on for the increasing meanings of modulus = mod $P_n\#$. with: $P_n\#$ - primorial. $P_{(\dots)} < P_0 < P_1 < P_2 \dots < P_n$ are the consequent primes. $C_{1\text{-}2\text{-}3\text{-}4}$ are primes and residues of mod$(P_n\#)$ $R_{4\text{-}3\text{-}2} = (C_{4\text{-}3\text{-}2} - C_1)$ that is length of the group = $(R_{4\text{-}3\text{-}2}-2)/2$ of odd numbers.			

Table 14.
The loopback of prime's groups and residues rearrangement according to the increasing modulus. With max R = const from mod($P_1\#$) to mod($P_3\#$).

from 1 to $P_2\#$ and represented in 1 line of **Table 2** where $\varphi(P_2\#)$ of groups No 3 of mod($P_2\#$) are situated with length $R_3 \leq 2P_2$ of the whole numbers. In this number, the two maximally long groups No 3 of mod($P_2\#$) with length max $R_3 = 2P_2$ of the whole numbers are represented. Then on the P.I. section from 1 to $P_3\#$ (at P_3 lines of **Table 2**), we'll get $\varphi(P_2\#)$ columns of group No 3 of mod($P_2\#$), with length $R_3 \leq 2P_2$, (P_3 lines in columns of groups No 3).

It is quite obvious, that in Section 9.1, it is proved that if by number P_3, "eliminate," that is to change for the model mod($P_3\#$) residue C_2 in the column of every $\varphi(P_2\#)$ group No 3 of mod($P_2\#$), then at the P.I. section from 1 to $P_3\#$, that is at the fractal $P_3\#$, we'll get $\varphi(P_2\#)$ groups No 2 of mod($P_3\#$) of the same length, that is $R_3 \leq 2P_2$ mod($P_2\#$), would become $R_2 \leq 2P_2$ of mod($P_3\#$) with changing the structure of alternance from $\leq P_2$ to $\leq P_3$.

As any eliminated residue C_2, during the rearrangement of groups from No 3 to No 2 for the mod($P_3\#$) cannot change the length on no subgroup that is all R_2 would permanently $\leq 2P_2$, and included in $\varphi(P_2\#)$ groups No 3 of mod($P_2\#$) there are two uncial, repeated P_3 times maximally long subgroups No 3 of mod($P_2\#$) with the alternance $\leq P_2$ with length maximal $R_3 = 2P_2$, that would be rearranged into two maximally long line-symmetrical groups No 2 of mod($P_3\#$) by "eliminating" the residues C_2 with the number multiple P_3, (1 time in P_3 lines). As all the other $\varphi(P_2\#)$–2 subgroups, rearranged from No 3 to No 2 for the mod($P_3\#$), are shorter than (P_2–1) of the off numbers, that is: $R_2 < 2P_2$, and in Sections 9.1 and 9.3, it is proved that there are no other ways of comparing or rearranging of the subgroups No 2 of mod($P_2\#$) with length $R_3 > 2P_2$. The order, type, and formula of indexing of two subgroups No 2 according to the increasing modulus are represented in **Table 13** b, c, d.

The length of these two line-symmetrical subgroups No 2 of mod($P_3\#$), the length of alternance $\leq P_3$, from C_1 to C_2, that is max $R_2 = (C_2–C_1) = (nP_1\# + P_2)–(nP_1\#–P_2) = 2P_2$; ($P_2$–1) of odd numbers. Two of these subgroups No 2 are situated within the P.I. section ($nP_1\#. \pm P_3$) with length $(C_B–C_A) = 2P_3$; (P_3–1) of odd numbers from C_A to C_B. The numerical values of these maximal groups No 2 is defined according to the formula: (multiple P_2 and multiple P_3) is ($nP_1\#. \pm .1$) and $(P_2P_3–n)P_1\#. \pm .1.$, with n and $(P_2P_3–n)$ define the line number for the group No 2 of mod($P_3\#$) in the column -$P_2{}^*P_3$ of the duplication of the max group No 4 of mod ($P_1\#$) with the period = $P_1\#$ (**Table 1**). That is n = (multiple $P_2{}^*P_3. \pm 1)/P_1\#$ = the whole < $P_2{}^*P_3/2$.

Herewith, it is quite obvious, and proved in Section 9.3, that all the other subgroups No 2 of mod($P_3\#$), with different lengths, rearranges from groups No 3 would be within the P.I., with length not exceeding the limit $=2P_3$ of the wholes.

And so on, for every of all posterior primes = P_n, at the increasing fractals $P_n\#$, with n is the whole, represented in **Tables 11** and **14** (the proof is indicated in Section 10) (numerical illustrations are in **Table 16**).

8. The loopback of rearrangement for $\varphi(P_n\#)$ groups No 3 (with 3 residues) from mod($P_{n-1}\#$) for mod($P_n\#$)

The loopback order of rearrangement of $\varphi(P_n\#)$ groups No 3, according to the increasing modulus, that are represented in column No 3 of **Table 14** at Section 8 are examined by steps, for every recurrent increasing fractal -$P_n\#$:

During the transition from mod($P_1\#$) to mod($P_2\#$) of the fractal -$P_1\#$ (1 line of **Table 1**) and every from $\varphi(P_1\#)$ groups No 4-3-2 of mod($P_1\#$) are repeated P_2 times. That at the P.I. section from 1 to $P_2\#$, we'll get $\varphi(P_1\#)$ columns of No 4-3-2 groups (P_2 limes at the column). Number P_2 according to diagonals of P_2 lines

for the fractal −*7#* **= 210, mod(7#), maxR$_3$ = 2*7, at the P.I. section (C_B–C_A) = 2*11, n = 3**								
C_A = 79 **$_AC$ = 109**	**…**	**C_1 = 83.** **$_1C$ = 113**	**5,3**	**C_2 = 89 и (7*13)** **(7*17) и $_2C$ = 121**	**3,5**	**C_3 = 97** **$_3C$ = 127**	**..**	**C_B = 101** **$_BC$ = 131**
for the fractal −*11#* = 2310, mod(*11#*), maxR$_3$ = 2*11, at the P.I. section (C_B–C_A) = 2*13, n = 1								
.C_A = 197. =n7#–*13* .$_AC$ = 2087. =11#–C_B	…	..C_1 = 199. =n7#–*11* $_1C$ = 2089. =11#–C_3	3,7, 5,3	(11*19),C_2 = 211 =n7#. ± .*1* $_2C$ = 2099,(11*191) =(11–n)7# ± 1	3,5, 7,3,	..C_3 = 221… =n7# + 11 …$_3C$ = 2111 (11–n)7# + 11	..	.C_B = 223 =n7# + 13 $_BC$ = 2113. (11–n)7# + 13
for the fractal −*13#* = 30,030, mod(*13#*), maxR$_3$ = 2*13, at the P.I. section (C_B–C_A) = 2*17, n = 3								
.C_A = 6913. =n11#–*17* .$_AC$ = 23,083. =13#–C_B	. …	..C_1 = 6917. =n11#–*13* $_1C$ = 23,087. =13#–C_3	11, 3,7, 5,3	(13*533),C_2 = 6931 =n11#. ± .*1* $_2C$ = 23,099,(13*1777) =(13–n)11# ± 1	3,5, 7,3, 11,	C_3 = 6943 =n11# + 13 …$_3C$ = 23,113 (13–n)11# + 13	..	.C_B = 6947 =n11# + 17 $_BC$ = 23,117. (13–n)11# + 17
for the fractal −*17#* = 510,510,mod(*17#*), maxR$_3$ = 2*17, at the P.I. section (C_B–C_A) = 2*19,n = 2								
.C_A = 60,041. =n13#–*19* .$_AC$ = 450,431. =17#–C_B	. …	..C_1 = 60,043. =n13#–*17* $_1C$ = 450,433. =17#–C_3	13, 11, 3,7, 5,3	C_2 = 60,059,(17*3533) =n13#. ± .*1* (17*26497), $_2C$ = 450,451 =(17–n)13# ± 1	3,5, 7,3, 11, 13,	C_3 = 60,077.. =n13# + 17 …$_3C$ = 450,467 (17-n)13# + 17	..	.C_B = 60,079 =n13# + 19 $_BC$ = 450,469 (17–n)13# + 19
for the fractal −*19#* = 9,699,690,mod(*19#*), maxR$_3$ = 2*19, at the P.I. section (C_B –C_A) = 2*23, n = 1								
C_A = 510,487 $_AC$ = 9,189,157	…	C_1 = 510,491 $_1C$ = 9,189,161	17.. 5,3	C_2 = 510,509 и кр.19. Кр.19 и $_2C$ = 9,189,181	3,5, …17,	..C_3 = 510,529 …$_3C$ = 9,189,199	..	.C_B = 510,533 $_BC$ = 9,189,203

Table 15.
The numerical examples of the two line-symmetrical maximally long subgroups No 3 (containing 3 residues = $C_{1\text{-}2\text{-}3}$), according to the increasing modulus.

"eliminates," that is rearranges to the mod($P_2\#$) one time every of P_2 repeated numbers of the P.I. section from 1 to $P_1\#$ (1 number at every P_2 line of every column No 4 and No 3), "eliminating" the residues C_1-C_2-C_3-C_4 in the groups No 4 (consult Section 8.1), and ALL numbers, besides the residues C_1-C_2-C_3 in the groups No 3 (consult Section 8.3).

8.1

It is quite obvious, that after "elimination" from every $\varphi(P_1\#)$ column of the group No 4 of mod($P_1\#$) one time every residue C_2 and C_3, we'll get at P_2 lines of every column of the No 4 groups of mod($P_1\#$), –TWO groups No 3 of mod($P_2\#$)

That is, we'll get $2\varphi(P_1\#)$ subgroups No 3 of mod($P_2\#$) with invariance length of the previous groups, that is R_4 $\varphi(P_1\#)$ groups of No 4 mod($P_1\#$), would become = R_3 for $2\varphi(P_1\#)$ groups No 3 of mod($P_2\#$), with changing the alternance structure from $\leq P_1$ to $\leq P_2$, that are situated at the P.I section from 1 to $P_2\#$ that is at the fractal-$P_2\#$.

Herewith within the $2\varphi(P_1\#)$ groups No 3 of mod($P_2\#$), there are accounted all residues = C_1 and = C_4 of mod($P_1\#$) and alternances $\leq P_2$ of such rearranged groups from No 4 of mod($P_1\#$) to No 3 of mod($P_2\#$). As along of P_2 duplication of the three adjoined groups No 4 of mod($P_1\#$), represented in **Table 3**, we'll get **Table 17**, with every residue = C_2 or = C_3, situated at one of the 3 lines of **Table 3** (for example, at line a of **Table 17**), is accounted as residue C_1 or C_4 at the other two adjoined groups No 4 (at lines: b or c of **Table 17**), where they are situated within 4 consequent residues, going as the second or the third, that is they are "excluded" as $C_{2 \text{ or } 3}$ in these adjoined groups (lines):

- for line (b) C_2, C_3. (C_4 = multiple to P_2), C_3 we'll get the alternance $\leq P_2$. $R_3 < 2P_2$.

- for line (c) C_2, (C_1 = multiple to P_2), C_2, C_3 we'll get the alternance $\leq P_2$. $R_3 < 2P_2$.

Thus, "eliminating" that means transition to mod($P_2\#$), one time every 4 residue in $\varphi(P_1\#)$ groups No 4 of mod($P_1\#$), at the P.I. sections from 1 to $P_2\#$, that is at the fractal -$P_2\#$, we'll, get the loopback, represented as $2\varphi(P_1\#)$ groups No 3 of mod ($P_2\#$) type: C_1 $_{pp}C_2$ $_{pp}$($C_{2\text{–}3}$ = multiple P_2) $_{pp}C_3$, with the alternances $\leq P_2$, length $R_3 \leq 2P_2$. Including, pure 2 subgroups No 3 of mod($P_2\#$) with length maximal $R_3 = 2P_2$.

8.2

Herewith, it is quite obvious that every three consequent residues of every subgroup No 3 according to the increasing group of mod($P_n\#$), represented in **Table 17**, are still within the P.I. section with length not exceeding - $2P_{n+1}$ of the whole numbers, as "eliminated" residues $C_{2;3}$ and $C_{1;4}$ of mod($P_n\#$) doesn't change the location of every subgroup No 3. That is, we'll get at the three adjoined groups No 3 lines of **Table 17** for the mod($P_2\#$): type (a) =(C_B–C_A) < $2P_3$; type (b) = (C_2–C_1) < $2P_3$; type (c) =(C_4–C_3) < 2_3.

Including pure two subgroups No 3 max $R_3 = 2P_2$, located within the maximally long section with length = $2P_3$ of the whole numbers. (The rearrangement is studied in Section 6).

8.3

Within the P_2 duplications $\varphi(P_1\#)$ of the groups No 3 of mod($P_1\#$), number P_2 "eliminated," that is rearranges to the mod($P_2\#$) 1 time every of all previously eliminated numbers of group No 3, except three residues C_1–C_2–C_3. That is, transits to the mod($P_2\#$) (P_2–3) of No 3 groups in every $\varphi(P_1\#)$ column of No 3 groups.

Then at the P.I. section from 1 to $P_2\#$, (that is included into fractal -$P_2\#$), we'll get the loop back, represented as (P_2–3) $\varphi(P_1\#)$ of No 3 groups repetition for the mod(P_2#) with "changing" the alternance from $\leq P_1$ to $\leq P_2$. Without changes the groups No 3 length for the mod(P_2#), R_3 = const and numbers composition within the alternances $\leq P_2$; that is the previously "eliminated" $\leq P_1$, according to the 1 least >1 from the number are accounted. Type: $C_1\ {}_{pp}C_2\ {}_{pp}C_3$.

With: R_3 = const =? (with to mod(P_2#) $R_3 << R_4 = 2P_2$).

8.4

Total at the fractal -P_2# at the P.I. section from 1 to P_2# we'll get the loopback of the rearranged groups No 3 for the mod(P_2#) represented in Sections 8.1 and 8.3.

That is: $2\varphi(P_1\#)$ (Section 8.1 with the length = $R_3 \leq 2P_2$) + $\varphi(P_1\#)(P_2{-}3)$ (Section 8.3 with the length $R_3 << R_4 = 2P_2$) = $P_2\varphi(P_1\#){-}3\varphi(P_1\#) + 2\varphi(P_1\#) = P_2\ \varphi(P_1\#){-}\varphi(P_1\#) = \varphi(P_1\#)(P_2{-}1) = \varphi(P_2\#)$ of the subgroups No 3 for the mod(P_2#), included at the alternances $\leq P_2$ with length $R_3 \leq 2P_2$, including two maximally long subgroups No 3 max $R_3 = 2P_2$.

8.5

As all $\varphi(P_2\#)$ of the subgroups No 3 of mod(P_2#) are examined in Sections 8.1 and 8.3, with all eliminated one time residues C_2 or C_3, examined in Section 8.1, cannot change the length = $R_3 \leq 2P_2$, none of the $2\varphi(P_1\#)$ groups, rearranged from No 4 to No 3 for the mod(P_2#). Herewith the residues C_1 and C_4 are also accounted in two adjoining groups of **Table 17** as C_2 or C_3. And indiscriminately $\varphi(P_1\#)(P_2{-}3)$ groups No 3 for the mod(P_2#), examined in Section 8.3 are shorter than limit $R_3 = 2P_2$.

So, there are no other ways to make groups No 3 of mod(P_2#) with length $R_3 > 2P_2$, besides the way to form the maximally long subgroups No 3 with length max $R_3 = 2P_2$, represented in Sections 6 and 8.1.

8.6

Thus we got, that for every recurrent = P_2, $\varphi(P_2\#)$ residues of mod(P_2#) situated in the fractal -P_2# (P_2 lines of **Table 1**), represented as loop back $\varphi(P_2\#)$ of subgroups No 3 (3 residues of mod(P_2#), represented in Section 8.4), that is pure THREE consequent prime ($_{A,B,C}C$) type: $P_2 < ({}_AC\text{-}{}_BC\text{-}{}_CC) < P_3^2$ at the P.I. section from P_2 to P_3^2. And further, from P_3^2 to P_2#, pure THREE consequent residue of mod(P_2#) at every P.I. section, with length not exceeding $2P_3$ of the whole numbers (see Section 8.2), with length of every subgroup No 3 at every section is $R_3 = ({}_CC{-}{}_AC) \leq 2P_2$ (consult Sections 8.1 and 8.4).

And so on, for every of all eventual primes = P_n, represented as the loopback of groups distribution of the residues No 3 at the increasing fractals -P_n# according to the increasing meanings of modulus-mod(P_n#) that proves the validity of section (b) of the Theorem 1 (loopback of groups No 3 is represented in column No 3 of **Table 14**).

9. The loopback of rearrangement for $\varphi(P_n\#)$ groups No 2 (2 residues) from the mod(P_{n-1}#) to mod(P_n#)

The looped back order of rearrangement $\varphi(P_n\#)$ of No 2 groups, according to the increasing modulus, that are represented in column No 4 of **Table 14**, in Section 9 are examined by steps for every recurrent increasing fractal -P_n#:

Representing the first P_2 lines in **Table 1** as one line, we'll get the fractal -P_2# according to mod(P_2#)-I.R.S of mod(P_2#) at the P.I. sections from 1 to P_2# (1 line of **Table 2**).

With $\varphi(P_2\#)$ line-symmetrical the least residues of mod(P_2#), which according to Section 2 are indexed according to $\varphi(P_2\#)$ groups of residues No 4-3-2 for the mod(P_2#).

				for the fractal - *7#* = 210, mod(*7#*), maxR_2 = 2*5, section (C_B–C_A) = 2*7, n = 1				
C_A = –1 $_AC$ = 197	...	C_1 = 1. $_1C$ = 199	3.	(5) и (7) (7*29) и (5*41)	3.	C_2 = 11 $_2C$ = 209	..	.C_B = 13 $_BC$ = 211
				for the fractal - *11#* = 2310, mod(*11#*), maxR_2 = 2*7, section (C_B–C_A) = 2*11, n = 4				
.C_A = 109. =n5#–*11* .$_AC$ = 2179. =11#–C_B	. ...	..C_1 = 113. =n5#–7 $_1C$ = 2183. =11#–C_2	5, 3.	(7*17),(11*11) =n5#. ± .*1* = 120 ± 1 (11*191),(7*313) =(7*11–n)5# ± 1	3, 5.	..C_2 = 127 ... =n5# + 7 ... $_2C$ = 2197 ... (7*11–n)5# + 7	..	.C_B = 131 =n5# + 11 $_BC$ = 2201. (7*11–n)5# + 11
				for the fractal - *13#* = 30,030, mod(*13#*), maxR_2 = 2*11, section (C_B–C_A) = 2*13, n = 45				
.C_A = 9437. =n7#–*13* .$_AC$ = 20,567. =13#–C_B	. ...	..C_1 = 9439. =n7#–*11* $_1C$ = 20,569. =13#–C_2	3, 7, 5. 3.	(11*859),(13*727) =n7#. ± .*1* = *9450* ± *1* (13*1583),(11*1871) =(11*13–n)7# ± 1	3, 5, 7, 3.	... C_2 = 9461 =n7# + 11 ... $_2C$ = 20,591.. (11*13-n)7# + 11	..	.C_B = 9463 =n7# + 13 $_BC$ = 20,593.. (11*13–n)7# + 13
				for the fractal - *17#* = 510,510,mod(*17#*), maxR_2 = 2*13, section (C_B–C_A) = 2*17,n = 94				
.C_A = 217,123. =n11#–*17* .$_AC$ = 293,353. =17#–C_B	. ...	..C_1 = 217,127. =n11#–*13* $_1C$ = 293,357. =17#–C_2	11, 3, 7, 5, 3.	(13*16703),(17*12773) =n11#. ± .*1* = *217,140*± *1* (17*17257),(13*22567) =(13*17–n)11# ± 1	3 5, 7, 3, 11,	..C_2 = 217,153.. =n11# + 13 ... $_2C$ = 293,383 (13*17–n)11# + 13	..	.C_B = 217,157 =n11# + 17 $_BC$ = 293,387 (13*17–n)11# + 17
				for the fractal - *19#* = 9,699,690,mod(*19#*), maxR_2 = 2*17, section (C_B –C_A) = 2*19, n = 2				
C_A = 60,041 $_AC$ = 9,639,611	...	C_1 = 60,043 $_1C$ = 9,639,613	13 3	кр.19., кр.17 = 60,061 Кр.17.,кр.19 = 9,639,631	3, .13.	..C_2 = 60,077 $_2C$ = 9,639,647	..	.C_B = 60,079 $_BC$ = 9,639,649

Table 16.
The numerical examples of the two line-symmetrical maximally long subgroups No 2 (containing 2 residues), according to the increasing modulus.

Type-(a) group from 4 to No 3 (C_1–C_4)	**C_A**	**C_1 (C_2 = mtp.P_2) *или* (C_3 = mtp.P_2), C_4**	**C_B**	**with R_4 is R_3 (C_4–C_1) < $2P_2$**	**at the P.I. section (C_B–C_A) < $2P_3$**
Type-(b) group from 4 to No 3 (C_2–C_3)	C_1	C_2, C_3, (C_4 = multiple to P_2),C_3	C_2	with R_4 is R_3 (C_3–C_2) < $2P_2$	at the P.I. section (C_2–C_1) < $2P_3$
Type-(c) group from 4 to No 3 (C_2–C_3)	C_3	C_2, (C_1 = multiple to P_2) C_2,C_3	C_4	with R_4 is R_3 (C_3–C_2) < $2P_2$	at the P.I. section (C_4 – C_3) < $2P_3$

Table 17.
The representation of rearrangement of every 3 adjoined subgroups No 4 of mod(P_1#) (represented in ***Table** 3), while P_2 duplication of fractal -P_1# from No 4 groups of mod(P_1#) to No 3 groups of mod(P_2#).*

At the transition from mod(P_2#) to mod(P_3#), the fractal -P_2# and every from the φ(P_2#) groups No 4-3-2 mod(P_2#) are repeated P_3 times. Then at the P.I. section from 1 to P_3# we'll get φ(P_2#) columns of No 4-3-2 groups (P_3 lines at the column). Number P_3 according to diagonals P_3 lines "eliminates," that is transits for the mod (P_3#) one time every P_3 of the duplicated numbers of the P.I. section from 1 to P_2#. (One number at every P_3 line of every column No 3 and No 2), "eliminating" the residues C_1-C_2-C_3 at the groups No 3 (consult Section 9.1) and ALL numbers, except the residues C_1-C_2 at the groups No 2 (consult Section 9.3).

9.1

It is quite obvious, that after "elimination" in every φ(P_2#) column of group No 3 of mod(P_2#), 1 time the residue - C_2, we'll get in P_3 line of every column of groups No 3 of mod(P_2#), one group No 2 of mod(P_3#), that is totally we'll get φ(P_2#) subgroups No 2 of mod(P_3#) with invariance length of previous groups, that is R_3 φ(P_2#) groups No 3 of mod(P_2#) would become = R_2 for φ(P_2#) groups No 2 of mod(P_3#) with changing the alternance composition from ≤P_2 to ≤P_3, that are situated at P.I. section from 1 to P_3# that is at the fractal -P_3#

Herewith in φ(P_2#) groups No 2 of mod(P_3#) are accounted all residues = C_1 and = C_3 of mod(P_2#) and alternances ≤P_3 of such "rearranged" groups from No 3 of mod(P_2#) to No 2 of mod(P_3#). As in the course of P_3 duplication of two adjoined groups No 3 of mod(P_2#), represented in **Table 5**, we'll get the table No 11 with each of the residues = C_2, situated on one of two lines of **Table 5** (for example, the line (a) **Table 18**, is accounted as the residue C_1 or C_3 at the other adjoined group No 3 (in line b). of **Table 12**), where it is represented in 3 consequent residues as the second one, that is "excluded" as C_2 at this adjoined group (line):

-for line (b) $_2$C., ($_3$C = multiple P_3)., C_2, we'll get the alternance ≤P_3. R_2 < $2P_2$

Thus, after "elimination" that is transferring to mod(P_3#), one time for every of 3 residues at φ(P_2#) groups No 3 of mod(P_2#), at the P.I. sections from 1 to P_3#, that is in fractal -P_3#, we'll get the loop back in the form of φ(P_2#) groups No 2 of mod(P_3#), type: C_1 $_{pp}$(C_2 = multiple P_2) $_{pp}$$C_3$, with the alternances ≤$P_3$, with length $R_2 \leq 2P_2$. Including pure 2 subgroups No 2 of mod(P_3#) with length maximal $R_2 = 2P_2$.

9.2

Herewith it is quite obvious that any two consequent residues of any subgroup No 2 according to the increasing mod(P_n#), represented in **Table 18** are still within the P.I. section with length not exceeding - $2P_{n+1}$ of the whole numbers, as the "eliminated" residues C_2 and $C_{1;3}$ of mod(P_n#) doesn't change the location of any subgroup No 2. That is, in two adjoined groups No 2 lines of **Table 18** for mod(P_3#) we get: type (a) =(C_B–C_A) < $2P_3$; type (b) =(C_1–$_1$C) < $2P_3$.

Including pure two subgroups No 2 max $R_2 = 2P_2$, located within the maximally long section with length = $2P_3$ of the whole numbers, two rearrangement is studies in Section 7.

9.3

Along with P_3 duplications $\varphi(P_2\#)$ of groups No 2 of mod($P_2\#$), the number P_3 "eliminates" that is transits to the mod($P_3\#$) 1 time every of all previously eliminated numbers of every group No 2, except two residues C_1–C_2. That it, it transits to mod($P_3\#$) (P_3–2) of groups No 2 in every $\varphi(P_2\#)$ column of No 2 groups.

Then at the P.I. section from 1 to $P_3\#$ that is within the fractal -$P_3\#$ we'll get the loopback, represented as $(P_3-2)\varphi(P_2\#)$ duplications of No 2 groups for the mod ($P_3\#$) with the alternance "changes" from $\leq P_2$ to $\leq P_3$. Without changing the length of groups No 2 for mod($P_3\#$), R_2 = const and numbers composition at the alternances $\leq P_3$ (as previously eliminated $\leq P_2$, for the 1 least >1 from the number is accounted). Type: C_1 $_{pppp}C_2$. With: R_2 = const =? (with to mod($P_3\#$) $R_2 < < R_3 = 2P_2$).

9.4

Totally at the fractal $P_3\#$ at P.I. section from 1 to $P_3\#$ we'll get the loopback of the rearranged groups No 2 for mod($P_3\#$) represented in Sections 9.1 and 9.3. That is: $\varphi(P_2\#)$ (Section 9.1 with length = $R_2 \leq 2P_2$) + $\varphi(P_2\#)(P_3-2)$ (Section 9.3 with length $R_2 < < R_3 = 2P_2$) = $P_3\varphi(P_2\#) - 2\varphi(P_2\#) + \varphi(P_2\#) = P_3\,\varphi(P_2\#) - \varphi(P_2\#) = \varphi(P_2\#)(P_3-1) = \varphi(P_3\#)$ of the subgroups No 2 for mod($P_3\#$), located within the alternances $\leq P_3$ with length $R_2 \leq 2P_2$, including two maximally long subgroups No 2 max $R_2 = 2P_2$.

9.5

As all $\varphi(P_3\#)$ of the subgroups No 2 of mod($P_3\#$) are examines in Sections 9.1 and 9.3, with every eliminated one time in the column residue C_2 examined in Section 9.1, can change the length = $R_2 \leq 2P_2$, of none of $\varphi(P_2\#)$ groups, rearranged from No 3 to No 2 for mod($P_3\#$), herewith residues C_1 and C_3 are excluded at the adjoined group of **Table 18** as C_2 and indiscriminately $(P_3-2)\varphi(P_2\#)$ groups No 2 of mod($P_2\#$), examined in Section 9.3 are shorter than limit R = $2P_2$.

Thus, there are no other ways of making groups No 2 of mod($P_3\#$) with length $R_2 > 2P_2$, but constructing two maximally long subgroups No 2 with length max $R_2 = 2P_2$, as examined in Sections 7 and 9.1.

9.6

And so we get, that for every recurrent prime = P_3, $\varphi(P_3\#)$ residues of mod ($P_3\#$) are in the fractal -$P_3\#$ (in P_3 lines of **Table 2**), represented as loopback $\varphi(P_3\#)$ of the subgroups No 2 (2 residues of mod($P_3\#$) are indicated in Section 9.4). Thus, pure TWO consequent primes ($_{A,B}C$) of type: $P_1, P_2 < (_AC - _BC) < P_3^2$ at the P.I. section from P_2 to P_3^2, and further, from P_3^2 to $P_3\#$ pure TWO consequent residues of mod ($P_3\#$), at every P.I. sections with length not exceeding $2P_3$ of the whole numbers (consult Section 9.2), with length of every subgroup No 2 at every section is: $R_2 = (_BC - _AC) \leq \mathbf{2}P_2$ (consult Sections 9.1 and 9.4).

And so on, for every from all eventual primes = P_n, in the form of loopback of residues of groups No 2 distribution at the increasing fractals -$P_n\#$, with the increasing values of modulus of mod($P_n\#$), that proofs the validity of section (c) of Theorem 1 (loopback of groups No 2 is represented in column No 4 of **Table 14**).

Type-(a) group from 3 to No 2 (C_1–C_3)	**C_A**	**C_1, (C_2 = multiple P_3)., C_3**	**C_B**	**with R_3 is R_2 (C_3–C_1) < $2P_2$**	**at the P.I. section (C_B–C_A) < $2P_3$**
Type (a) group from 3 to No 2 ($_2C$–C_2)	$_1C$	$_2C$., ($_3C_1$ = multiple P_3).,C_2	C_1	with R_3 is R_2 (C_2–$_2C$) < $2P_2$	At the P.I. section (C_1–$_1C$) < $2P_3$

Table 18.
*Representation of rearrangement of any 2 adjoined subgroups No 3 of mod($P_2\#$) (represented in **Table 5**), within P_3 duplications of fractal -$P_2\#$. From groups No 3 of mod ($P_2\#$) to the groups No 2 of mod ($P_3\#$).*

10. Proof of section (b) and section (c) of Theorem 1

While examining the P.I., represented in the form of alternance (array) of primes (according to the 1 least prime factor > 1 from every whole number), we'll get that for every recurrent prime -P_n; the P.I. is the line-symmetrical primary-repeated fractal -$P_n\#$, located at the P.I. section from 1 to $P_n\#$, represented as $\varphi(P_n\#)$ of the I.R.S. residue of mod($P_n\#$), between which the P.I. sections are situated (with different length), represented as the alternances (array) of different amounts of different first primes $\leq P_n$ – NOT residues mod($P_n\#$).

By indexing $\varphi(P_n\#)$ of the least residues of every recurrent fractal -$P_n\#$, groups pure with 4, 3, and 2 consequent residue of mod($P_n\#$) (analogously as in Section 2). We'll get that every recurrent fractal -$P_n\#$ has got three types of arrays of the subgroups of residues of mod($P_n\#$): with $\varphi(P_n\#)$ groups: No 4 (with 4 residues), No 3 (with 3 residues), and No 2 (with 2 residues, repeated without changes with the period = $P_n\#$).

At every recurrent transition, for example, from mod($P_2\#$) to mod($P_3\#$), the first line of fractal -$P_2\#$ and group No 4-3-2 of mod($P_2\#$) are repeated P_3 times (consult P_3 lines of **Table 2**). At the P.I. section from 1 to $P_3\#$ we'll get $\varphi(P_2\#)$ columns of groups No 4-3-2 of mod($P_2\#$) within the alternances $\leq P_2$ (P_3 lines in column), "eliminating" 1 number multiple to -P_3 (in line of every column No 4-3-2), that is by transition of this group to mod($P_3\#$) with changing of its length: from $R_{(4,3)}$ to $R_{3,2}$ and the alternance composition from $\leq P_2$ to $\leq P_3$, we'll get $\varphi(P_3\#)$ groups No 4-3-2 of mod($P_3\#$).

The rearrangement order of these subgroups No 4-3-2 at the increasing modulus is proved in Sections 5, 6, 7, 8, and 9. Representing as one line of the P.I. section from 1 to $P_3\#$ we'll get the fractal -$P_3\#$ of mod($P_3\#$), with $\varphi(P_3\#)$ groups No 4-3-2 of mod($P_3\#$).

And so on for every recurrent prime = P_n, the results of proof are demonstrated in Sections from 5 to 9 and for visualization they are grouped together in **Table 14**.

As far as we know that for every recurrent prime -$P_n = P_{(1)}$, the P.I. is the line-symmetrical primordial repeated fractal -$P_{(1)}\#$, located at the P.I. section from 1 to $P_{(1)}\#$. There are $\varphi(\boldsymbol{P}_{(1)}\#)$ residues of mod($\boldsymbol{P}_{(1)}\#$) between which are located the alternances (arrays) of primes $\leq P_{(1)}$ (1 the least>1 from every NOT residue of mod ($P_{(1)}\#$)).

According to Section 5, there is the only maximally long subgroup No 4 (with 4 residues) of mod($P_{(1)}\#$) with length $\max R_4 = 2P_{(2)}$.

Let us assume that there are such primes $P_{(2)}$ or $P_{(3)}$, for which at the P.I., represented as alternance $\leq P_{(2)}$, in the form of fractal $-P_{(2)}\#$ we can form more than two maximally long subgroups No 3 of mod($P_{(2)}\#$), with: (or) $R_3 > 2P_{(2)}$ or at the fractal $-P_{(3)}\#$ with P.I. is represented by alternances $\leq P_{(3)}$, we can compare more than maximally long subgroups No 2 of mod($P_{(3)}\#$), with: (or) $R_2 > 2P_{(2)}$.

Then, in the course of the opposite reduction of the modulus, that is at the result of $P_{(2)}{}^*P_{(3)}$ repetition of such subgroups as No 3 or No 2 with the repetition period $P_{(1)}\#$ (for the downward meanings of numbers) and backing up the $P_{(2)}$ and $P_{(3)}$ numbers as residues (according to the decreased moduli)**, that are situated in $P_{(2)}{}^*P_{(3)}$ lines analogues to **Table 1**. At the upper lines of such columns, consisting of $P_{(2)}{}^*P_{(3)}$ lines, we'll get the P.I as fractal-$P_{(1)}\#$, represented by alternances $\leq P_{(1)}$, where at the P.I. section from 1 to $P_{(1)}\#$ would be located more than one subgroup No 4 (with 4 residues) of mod($P_{(1)}\#$) or subgroups No 4, with $R_4 > 2P_{(2)}$. It is quite obvious, that any such group No 4 according to the reestablished mod($P_{(1)}\#$) would be line-symmetrical to the left and to the right from the symmetry center of number = $P_{(1)}\#/2$, that means formed by two different ways, that contradicts to the axiom set.

**Number $P_{(3)}$ of the fractal -$P_{(3)}$# -NOT residue of mod(P_3#), located in the group No 2 of mod($P_{(3)}$#) within the alternance $\leq P_{(3)}$ with length $R_{(2)} > 2P_{(2)}$, and for mod($P_{(2)}$#) would be accounted as the third residue in the group No 3 within the alternance $\leq P_{(2)}$, without changing the length of this group No 3 with $R_{(2)}$ is $R_3 > 2P_{(2)}$.

Number $P_{(2)}$ of the fractal -$P_{(2)}$# -NOT residue of mod(P_2#) is located in the group No 3 of mod($P_{(2)}$#) within the alternance $\leq P_{(2)}$, with length $R_3 > 2P_{(2)}$, and for mod

($P_{(1)}$#), would be accounted as the fourth residue in the group No 4 within the alternance $\leq P_{(1)}$, without changing the length of this group No 4 with $R_{(3)}$ is $R_4 > 2P_{(2)}$.

11. Conclusion

Thus, Theorem 1 allowed us to prove the existence of a new law in mathematics – "on the plume distribution of Prime numbers." Since the methods used in number theory do not allow us to approach the problem of the distribution of prime numbers, it means that further expansion of the method proposed in the article for studying the natural series of numbers will simplify and solve many other problems that are not solved in mathematics.

So from Theorem 1 "Loopback of primes distribution" follows:

Theorem No 2. For every whole number = N at the P.I. section from 1 to N + $2\sqrt{N}$:

1. Primes are located as groups, pure three consequent primes of (P_1-P_2-P_3) type. Herewith the distance from the first to the third prime of every group is less than $2\sqrt{N}$ of the whole numbers, that is $(P_3–P_1) < 2\sqrt{N}$ whole numbers.

2. The same primes are distributed as the loopback, pure two consequent primes at every P.I. sections, shorter than $2\sqrt{N}$ whole numbers.

Proof. Every whole number = N is located within the squared two consequent primes: $P_1^2 < N \leq P_2^2$ with: $2\sqrt{N} > 2P_1$.

That means every N is located within the fractal -P_1#. Then:

1. From the section (b) of the theorem "Loopback of primes distribution" follows, that at fractal -P_1# of mod(P_1#) at the P.I. section from 1 to $P_2^2 \geq (N + 2\sqrt{N})$, at every P.I. section with length not exceeding $2P_2$ of the whole numbers is located at the subgroup from three consequent primes of (P_1-P_2-P_3) type, with length of every subgroup, that is distance from the first to the third prime of every subgroup doesn't exceed $2P_1$ whole numbers, that is $(P_3–P_1) < 2P_1$ whole numbers. As length of every section $2\sqrt{N} >$ length of the section = $2P_1$. Then from 1 to N + $2\sqrt{N}$ – every:

$(P_3–P_1) < 2\sqrt{N}$ of the whole numbers.

2. From the section (c) of the Theorem "Loopback of primes distribution" it follows, that at the fractal -P_1# of mod(P_1#) at the P.I. section from 1 to $P_2^2 \geq (N + 2\sqrt{N})$ at every I.P. section with length not exceeding $2P_1$ of the

whole numbers, there is the subgroup from two consequent primes P_1 and P_2. As $2\sqrt{N}$ whole numbers $>2P_1$ whole numbers, that means, that at the P.I. section from 1 to N + $2\sqrt{N}$ at every P.I. section with length not exceeding $2\sqrt{N}$ whole numbers, there is the loopback of primes, represented as the subgroup for two consequent primes.

Genuinely: Every P.I. section with length = $2\sqrt{N}$ of the whole numbers is located at the fractal -P_1# at the P.I. section. $<P_2^2$ as, $P_1^2 < (N + 2\sqrt{N}) \leq P_2^2$, with: $2\sqrt{N} > 2P_1$.

It is feasible, that there is a P.I. section with length = $2\sqrt{N}$ of the whole numbers, where there are no two primes, that is, two consequent primes are located at the P.I. section with length exceeding – $2\sqrt{N}$ of the whole numbers $>2P_1$, but this contradicts to section (c) of the Theorem "Loopback of primes distribution," that states, that is every fractal -P_1# according to mod(P_1#), on every P.I. sections with length not exceeding **2**P_1 of the whole numbers, there is a subgroup No 2 with 2 residues of mod(P_1#), that is two primes $<P_2^2$.

Author details

Shcherbakov Aleksandr Gennadiyevich
Independent Researcher, Russia

*Address all correspondence to: ag_ask@mail.ru

References

[1] Bergman GN. Chislo i nauka o nem. [A Number and A Theory of Numbers]. Moscow: Gostekhizdat; 1949. [in Russian]

[2] Vinogradov IM. Osnovy teorii chisel. [Fundamentals of A Theory of Numbers]. Moscow: Gostekhizdat; 1952. [in Russian]

[3] Ozhigova YP. Razvitie teorii chisel v Rossii. [Development of A Theory of Numbers in Russia]. 2nd ed. Moscow: Directory of the Ukrainian SSR; Vol. 38. 2003. pp. 270-275

[4] Serpinsky V. Chto my znaem i chego ne znaem o prostyh chislah. [What Do We Know and What We Don't Know About Simple Numbers]. Moscow: State Publishing House for physical and mathematical Literature; 1963. [in Russian]

[5] Aleksandrov PS. Jenciklopedija jelementarnoj matematiki/ kniga pervaja arifmetika// [Encyclopedia of Elementary Mathematics. Volume 1 Arithmetic]. Izdatel'stvo tehnikoteoreticheskoj literatury; Moscow. 1951. pp. 255-282. [in Russian]

[6] Prahar K. Raspredelenie prostyh chisel. Izdatel'stvo"MIR"; Moscow. 1967. p. 511

[7] Shherbakov AG. O raspredelenii prostyh chisel. [On Distribution of Prime Numbers]. Vol. 10. Perspektivy nauki. - Tambov, Izdatel'stvo MOO Fond razvitija nauki i kul'tury; 2013. pp. S142-S147. [in Russian]

[8] Shcherbakov AG. Chetyre prostykh chisla mezhdu kvadratami dvukh posledovatelnykh prostykh chisel. In: NAUKA SEGODNYa globalnyye vyzovy i mekhanizmy razvitiya: materialy Mezhdunarodnoy nauchno-prakticheskoy konferentsii. Vologda: Disput; 2018. pp. 77-84

[9] Shherbakov AG. Dokazatel'stvo Gipotezy Brokara (O chetyreh prostyh chislah). In: Aspirant. Vol. 12. Prilozhenie k zhurnalu Vestnik ZBGU; 2018. pp. S.109-S.116

[10] Shherbakov AG. The Length Of An Interval Of A Positive Integers Sequence Represented (Measured) By An Alternation Of The First Prime Numbers. GJPAM. 2015;**11**(4): 1803-1818

Chapter 5

Moments of Catalan Triangle Numbers

Pedro J. Miana and Natalia Romero

Abstract

In this chapter, we consider the Catalan numbers, $C_n = \frac{1}{n+1}\binom{2n}{n}$, and two of their generalizations, Catalan triangle numbers, $B_{n,k}$ and $A_{n,k}$, for $n, k \in \mathbb{N}$. They are combinatorial numbers and present interesting properties as recursive formulae, generating functions and combinatorial interpretations. We treat the moments of these Catalan triangle numbers, i.e., with the following sums: $\sum_{k=1}^{n} k^m B_{n,k}^j$, $\sum_{k=1}^{n+1} (2k-1)^m A_{n,k}^j$, for $j, n \in \mathbb{N}$ and $m \in \mathbb{N} \cup \{0\}$. We present their closed expressions for some values of m and j. Alternating sums are also considered for particular powers. Other famous integer sequences are studied in Section 3, and its connection with Catalan triangle numbers are given in Section 4. Finally we conjecture some properties of divisibility of moments and alternating sums of powers in the last section.

Keywords: Catalan numbers, combinatorial identities, binomial coefficients, moments

1. Introduction

After the binomial coefficients, the well-known Catalan numbers $(C_n)_{n \geq 0}$ are the most frequently occurring combinatorial numbers. They are treated deeply in many books, monographs, and papers (e.g., [1–20]). Catalan numbers play an important role and have a major importance in computer science and combinatorics.

They appear in studying astonishingly many combinatorial problems. They count the number of different ways to triangulate a regular polygon with $n + 2$ sides; or, the number of ways that $2n$ people seat around a circular table are simultaneously shaking hands with another person at the table in such a way that none of the arms cross each other, and also in tree enumeration problem, see these examples and others in [19, 20].

Other applications of the Catalan numbers appear in engineering in the field of cryptography to form keys for secure transfer of information; in computational geometry, they are generally used in geometric modeling; they may be also found in geographic information systems, geodesy, or medicine.

There are several ways to define Catalan numbers; one of them is recursively by $C_0 = 1$ and $C_n = \sum_{i=0}^{n-1} C_i C_{n-1-i}$ for $n \geq 1$; the first terms in this sequence are

$$1,\ 1,\ 2,\ 5,\ 14,\ 42,\ 132,\ \ldots \tag{1}$$

The generating formula for Catalan numbers is

$$C(x) := \frac{1-\sqrt{1-4x}}{2x} = \sum_{n\geq 0} C_n x^n, \qquad 0 < x < 1/14 \tag{2}$$

[10] and ([20], Proposition 1.3.1).
Catalan triangle numbers $(B_{n,k})_{n,k\geq 1}$ and $(A_{n,k})_{n,k\geq 1}$ are defined by

$$B_{n,k} := \frac{k}{n}\binom{2n}{n-k}, \qquad A_{n,k} := \frac{2k-1}{2n+1}\binom{2n+1}{n+1-k} \; n,k\in\mathbb{N},\ \ k\leq n+1. \tag{3}$$

Notice that $B_{n,1} = A_{n,1} = C_n$. In [14], Shapiro introduced Catalan triangles whose entries are given by the coefficients

$$\sum_{n\geq k} B_{n,k} x^n = x^k C^{2k}(x), \tag{4}$$

see a more general approach in [10].
Although the numbers $B_{n,k}$ (and also $A_{n,k}$) are not as well-known as Catalan numbers, they have also several applications, for example, $B_{n,k}$ is the number of walks of n steps, each in direction N, S, W, or E, starting at the origin, remaining in the upper half-plane and ending at height k; see more details in [4, 13, 14, 16] for additional information.
Both Catalan triangle numbers may be written in unified expression. We consider combinatorial numbers $(C_{m,k})_{m\geq 1,k\geq 0}$, given by

$$C_{m,k} := \frac{m-2k}{m}\binom{m}{k}. \tag{5}$$

These combinatorial numbers $(C_{m,k})_{m\geq 1,k\geq 0}$ are suitable rearrangements of the known ballot numbers $(a_{m,k})$ with $a_{m,k} = \frac{k+1}{m+1}\binom{2m-k}{m}$ for $m \geq 0$ and $0 \leq k \leq m$, i.e.,

$$a_{m,k} = C_{2m+1-k,m-k}, \qquad C_{m,k} = a_{m-k-1,m-2k-1}, \tag{6}$$

see example [21]. Note that $C_{2n,n-k} = B_{n,k}$ and also $C_{2n+1,n+1-k} = A_{n,k}$. In ([9], Theorem 1.1), the authors show that any binomial coefficient can be written as weighted sums along the rows of the Catalan triangle, i.e.,

$$\binom{n+k+1}{k} = \sum_{j=0}^{k} C_{n,j} 2^{k-j}. \tag{7}$$

The generalized kth Catalan numbers ${}_kC_n := \frac{1}{n}\binom{nk}{n-1}$, $k \geq 1$, are presented in [17] to count the number of ways of subdividing a convex polygon into k disjoint $(n+1)$-polygons by means of nonintersecting diagonals, $k \geq 1$; see also [2, 11].

In this paper, our main objective is to study in detail the moments of Catalan triangle numbers:

$$\sum_{k=1}^{n} k^m B_{n,k}^j, \qquad \sum_{k=1}^{n+1} (2k-1)^m A_{n,k}^j, \tag{8}$$

for $j, n \in \mathbb{N}$ and $m \in \mathbb{N} \cup \{0\}$. In previous papers, the authors have considered some particular cases of these sums: for $j = 1$ and $m = 0$ in [14], for $j = 2$ in [12, 13], and for $j = 3$ and $m = 0$ in [22]. In [7], the authors solved a conjecture posed in [22] about divisibility properties in the case $m = 0$. However, there are no results in the literatures for moments for $j > 2$. We complete and present a full treatment of these moments, for $j = 1$ in Section 2 and for $j = 2$ and for some cases of $j = 3$ in Section 4.

In the paper [23], the authors treat several families of binomial sum identities whose definition involves the absolute value function. Here we present alternating sums of for several powers of Catalan triangle numbers (Theorem 2.2, Proposition 4.1 (iii), and Proposition 4.4 (iii)). In ([24], Theorem 2.3), the following identityis proved:

$$\sum_{k=1}^{n} (-1)^k k^2 B_{n,k}^2 = n(n-2)(2n-1)C_{n-1}, \qquad n \geq 1. \tag{9}$$

In this paper, we treat $\sum_{k=1}^{n}(-1)^k k^2 B_{n,k}^j$ and $\sum_{k=1}^{n+1}(-1)^k k^2 A_{n,k}^j$ for $j \in \{1, 2, 3, 4, 5\}$, and we conjecture some divisibility properties in Conjecture 5.7.

The WZ theory is a powerful tool to show hypergeometric identities. We have applied this tool in Theorem 2.1 to check certain identities. In detail, we have used the Maple program and the EKHAD package as software for the WZ method; see ([25], Example 7.5.3). Although analytic proofs are not presented, alternative proofs as to apply WZ theory [26, 27] or some mathematical software indicate us what these identities hold. Note that an analytic proof will give us some extra information about these natures of the sums.

In Section 3, we prove new identities involving sequences $(a(n))_{n \geq 0}$ and $(b(n))_{n \geq 1}$ where

$$a(n) := \sum_{k=0}^{n} \binom{n+k}{n}^2, \qquad b(n) := \sum_{k=0}^{n} \frac{n-k}{n} \binom{n-1+k}{n-1}^2 \qquad n \in \mathbb{N}, \tag{10}$$

and Catalan numbers $(C_n)_{n \geq 0}$. In Theorems 3.1 and 3.2, we show that for $n \geq 1$,

$$2(2n+1)a(n) - na(n-1) = (21n+8)\left(\frac{n+1}{2}\right)^2 C_n^2, \tag{11}$$

$$2(2n+1)b(n+1) - nb(n) = (7n^2+8n+2)C_n^2. \tag{12}$$

Lemma 3.3 shows that sequences $(a(n))_{n \geq 1}$ and $(b(n))_{n \geq 1}$ are deeply connected with Catalan numbers. Recurrence relations (30) and (36) (and polynomials in these relations) play delicate roles which allow to give proof of the identity:

$$((n+1)C_n)^2 = 4(3a(n-1) - 2b(n)), \qquad n \geq 1, \tag{13}$$

(Theorem 3.4).

In Section 4, we give the moments of second order in Theorem 4.2 and 4.3, and for third order, we present that

$$\sum_{k=0}^{n} B_{n,k}^3 = \frac{n+1}{2} C_n b(n), \qquad \sum_{k=1}^{n+1} A_{n,k}^3 = (n+1)C_n\Big(\big(2(n+1)C_n\big)^2 - 3a(n)\Big), \quad (14)$$

for $n \geq 1$; see also ([22], Section 3).

Finally, we conjecture some divisibility properties in Section 5; in particular

$$\sum_{k=1}^{n} k^{2m} B_{n,k} = \frac{n+1}{2} C_n n P_{m-1}(n), \quad (15)$$

$$\sum_{k=1}^{n} k^{2m-1} B_{n,k} = 2^{n-m-1} Q_{m-1}(n), \quad (16)$$

$$\sum_{k=1}^{n+1} k^{2m} A_{n,k} = (n+1) C_n R_{m-1}(n), \quad (17)$$

$$\sum_{k=1}^{n+1} k^{2m-1} A_{n,k} = 2^{2n} S_{m-1}(n), \quad (18)$$

where P_{m-1}, Q_{m-1}, R_{m-1} and S_{m-1} are polynomials of integer coefficients at the degree at most $m-1$ (Conjectures 5.1 and 5.2). In Conjecture 5.3, we state that the factor $\frac{n+1}{2}C_n$ could divide $\sum_{k=0}^{n} k^{2m} B_{n,k}^3$ for $m, n \in \mathbb{N}$; similarly the factor $(n+1)C_n$ might divide $\sum_{k=0}^{n+1} (2k-1)^{2m} A_{n,k}^3$ for $m, n \in \mathbb{N}$ (Conjecture 5.4). Similar conjectures about moments of fourth order and alternating sums are also presented in Conjectures 5.5–5.7.

2. Sums and alternating sums of Catalan triangle numbers

Catalan triangle numbers $(B_{n,k})_{n \geq 1, 1 \leq k \leq n}$ were introduced in [14]. These combinatorial numbers $B_{n,k}$ are the entries of the following Catalan triangle:

$n \backslash k$	1	2	3	4	5	6	...
1	1						
2	2	1					
3	5	4	1				
4	14	14	6	1			
5	42	48	27	8	1		
6	132	165	110	44	10	1	
...	...	...	...	...	...	...	...

(19)

which are given by

$$B_{n,k} := \frac{k}{n}\binom{2n}{n-k}, \quad n, k \in \mathbb{N}, \quad k \leq n. \quad (20)$$

Notice that $B_{n,1} = C_n$ and $B_{n,n} = 1$ $n \geq 1$.

In the last years, Catalan triangle (19) has been studied in detail. For instance, the formula

$$\sum_{k=1}^{i} B_{n,k} B_{n,n+k-i}(n+2k-i) = (n+1)C_n \binom{2(n-1)}{i-1}, \quad i \leq n, \quad (21)$$

which appears in a problem related with the dynamical behavior of a family of iterative processes has been proved in ([8], Theorem 5). These numbers $(B_{n,k})_{n\ge k\ge 1}$ have been analyzed in many ways. For instance, symmetric functions have been used in [1], recurrence relations in [15], or in [6] the Newton interpolation formula, which is applied to conclude divisibility properties of the sums of products of binomial coefficients.

Other combinatorial numbers $A_{n,k}$ defined as follows

$$A_{n,k} := \frac{2k-1}{2n+1}\binom{2n+1}{n+1-k}, \quad n,k\in\mathbb{N}, \;\; k\le n+1, \tag{22}$$

appear as the entries of this other Catalan triangle,

$n \backslash k$	1	2	3	4	5	6	...
1	1	1					
2	2	3	1				
3	5	9	5	1			
4	14	28	20	7	1		
5	42	90	75	35	9	1	
6	132	297	275	154	54	11	1
...	...	...	...	...	...	...	...

(23)

which is considered in [13]. Notice that $A_{n,1} = C_n$ and $C_{2n+1,n-k+1} = A_{n,k}$ for $k\le n+1$.

Entries $B_{n,k}$ and $A_{n,k}$ of the above two particular Catalan triangles satisfy the recurrence relations

$$B_{n,k} = B_{n-1,k-1} + 2B_{n-1,k} + B_{n-1,k+1}, \qquad k\ge 2, \tag{24}$$

and

$$A_{n,k} = A_{n-1,k-1} + 2A_{n-1,k} + A_{n-1,k+1}, \qquad k\ge 2. \tag{25}$$

For $m\in\mathbb{N}\cup\{0\}$, we define the moments of order m by the sum

$$\Delta_m(n) := \sum_{k=0}^{n} k^m B_{n,k}, \qquad \Lambda_m(n) := \sum_{k=0}^{n+1}(2k-1)^m A_{n,k}, \qquad n\ge 1. \tag{26}$$

As it was shown in [14], the values of the sums (or moments of order 0) of $B_{n,k}$ and $A_{n,k}$ are expressed in terms of Catalan numbers; see item (i) and (iii) in the next theorem. We apply the WZ theory to show the following moments for $m\in\{0,1,\ldots 7\}$.

Theorem 2.1. *For $n\in\mathbb{N}$, the following identities hold*:

i. $\Delta_0(n) = \frac{n+1}{2}C_n,$

$\Delta_2(n) = n\frac{n+1}{2}C_n,$

$\Delta_4(n) = n(2n-1)\frac{n+1}{2}C_n,$

$\Delta_6(n) = n\left(6n^2+4n+1\right)\frac{n+1}{2}C_n.$

ii. $\Delta_1(n) = 2^{2n-2}$,
$\Delta_3(n) = 2^{2n-3}(3n-1)$,
$\Delta_5(n) = 2^{2n-4}(15n(n-1)+2)$,
$\Delta_7(n) = 2^{2n-5}(105n^3-210n^2+147n-34)$.

iii. $\Lambda_0(n) = (n+1)C_n$,
$\Lambda_2(n) = (n+1)C_n(4n+1)$,
$\Lambda_4(n) = (n+1)C_n(32n^2+8n+1)$,
$\Lambda_6(n) = (n+1)C_n(384n^3-32n^2+12n+1)$.

iv. $\Lambda_1(n) = 2^{2n}$,
$\Lambda_3(n) = 2^{2n}(6n+1)$,
$\Lambda_5(n) = 2^{2n}(60n^2+1)$,
$\Lambda_7(n) = 2^{2n}(840n^3-420n^2+126n+1)$.

For alternating sums, the following theorem was proved in [5] and ([22], Corollary 1.3).

Theorem 2.2. *For $n \geq 1$, we have*

i. $\sum_{k=1}^{n}(-1)^k B_{n,k} = -C_{n-1}$,

ii. $\sum_{k=1}^{n+1}(-1)^k A_{n,k} = 0$.

Other interesting combinatorial numbers which have been deeply studied in the last decade are the well-known harmonic numbers $(H_n)_{n\geq 1}$. These numbers are given by the following formula:

$$H_n = \sum_{k=1}^{n}\frac{1}{k}, \quad n \in \mathbb{N}. \tag{27}$$

A deep treatment of closed formulas for the sums of the form $\sum_{k=1}^{n} a_k H_k$ is given in [18]. Also, the WZ theory is applied to get identities in [26], and infinite series involving harmonic numbers is presented in [3]. See other approaches in ([28], Chapter 7) and reference therein.

In ([22], Corollary 1.5) the next relationships between Catalan triangle numbers and harmonic numbers $(H_n)_{n\geq 1}$ are given.

Corollary 2.3. *For $n \geq 1$, we have*

i. $\sum_{k=0}^{n-1} B_{n,k}H_{n-k} = \frac{(2nH_n-1)(n+1)}{4n}C_n - \frac{2^{2n-1}-1}{2n}$,

ii. $\sum_{k=1}^{n} A_{n,k}H_{n-k+1} = H_n(n+1)C_n - \frac{2^{2n}-1}{2n+1}$.

Remark. It is worth to consider other powers of Catalan triangle numbers and harmonic numbers to obtain, for example, formulae of

$$\sum_{k=0}^{n-1} B_{n,k}^2 H_{n-k}, \qquad \text{and} \qquad \sum_{k=1}^{n} A_{n,k}^2 H_{n-k+1}. \tag{28}$$

3. Sums of squares of combinatorial numbers

We consider the sequence of integer numbers defined by

$$a(n) := \sum_{k=0}^{n} \binom{n+k}{n}^2, \qquad n \in \mathbb{N} \cup \{0\}. \tag{29}$$

Note that $a(0) = 1$, $a(1) = 5$, $a(2) = 46$, $a(3) = 517$, $a(4) = 6376$, etc. This sequence appears indexed in the On-Line Encyclopedia of Integer Sequences by N.J. A. Sloane [16] with the reference $A112029$. V. Kotesovec in 2012 proved the following recurrence relation:

$$p_1(n)a(n) = p_2(n)a(n-1) + p_3(n)a(n-2), \qquad n \geq 2, \tag{30}$$

where polynomials $(p_i)_{i \in \{1,2,3\}}$ are defined by

$$p_1(n) := 2(2n+1)(21n-13)n^2, \tag{31}$$

$$p_2(n) := 1365n^4 - 1517n^3 + 240n^2 + 216n - 64, \tag{32}$$

$$p_3(n) := -4(n-1)(2n-1)^2(21n+8). \tag{33}$$

Next, in the following theorem, we provide an identity which relates the square of Catalan numbers and $(a(n))_{n \geq 0}$.

Theorem 3.1. *For $n \geq 1$, the following identity holds*

$$2(2n+1)a(n) - na(n-1) = (21n+8)\binom{n+1}{2}^2 C_n^2. \tag{34}$$

Proof. We show this identity by induction method. For $n = 1$, we check directly that $29 = (21 \cdot 1 + 8)C_1^2$. Now suppose that the identity holds for any $m \leq n$. Note that

$$\begin{aligned}(21n+8)\binom{n+2}{2}^2 C_{n+1}^2 &= (21n+8)4(2n+1)^2\binom{n+1}{2}^2 C_n^2 \\ &= 4(2n+1)^2(2(2n+1)a(n) - na(n-1)),\end{aligned}$$

where we have applied the induction hypothesis. Then we apply the law of recurrence (30) to get that

$$\begin{aligned}&(21n+8)(21n+29)\binom{n+2}{2}^2 C_{n+1} \\ &= 8(21n+29)(2n+1)^3 a(n) + p_3(n+1)a(n-1) \\ &= p_1(n+1)a(n+1) + \left(8(21n+29)(2n+1)^3 - p_2(n+1)\right)a(n) \\ &= 2(2n+3)(21n+8)(n+1)^2 a(n+1) - (21n+8)(n+1)^3 a(n) \\ &= (21n+8)(n+1)^2(2(2n+3)a(n+1) - (n+1)a(n)),\end{aligned}$$

and we conclude the proof. □

Now we consider this second sequence of integer numbers defined by

$$b(n) := \sum_{k=0}^{n} \frac{k}{n} \binom{2n-k-1}{n-1}^2 = \sum_{k=0}^{n} \frac{n-k}{n} \binom{n-1+k}{n-1}^2, \qquad n \in \mathbb{N}. \tag{35}$$

Note that $b(1) = 1$, $b(2) = 3$, $b(3) = 19$, $b(4) = 163$, $b(5) = 1625$, etc. This sequence also appears indexed in the On-Line Encyclopedia of Integer Sequences by N.J.A. Sloane [16] with the reference $A183069$, and V. Kotesovec proved the following recurrence relation:

$$q_1(n)b(n) = q_2(n)b(n-1) + q_3(n)b(n-2), \qquad n \geq 3, \tag{36}$$

where polynomials $(q_i)_{i \in \{1,2,3\}}$ are defined by

$$q_1(n) := 2n^2(2n-1)\left(7n^2 - 20n + 14\right), \tag{37}$$

$$q_2(n) := 455n^5 - 2427n^4 + 4850n^3 - 4406n^2 + 1728n - 216, \tag{38}$$

$$q_3(n) := -4(n-2)(2n-3)^2\left(7n^2 - 6n + 1\right). \tag{39}$$

In a similar way, we obtain an identity which relates numbers $(b(n))_{n \geq 1}$ to the square of Catalan numbers.

Theorem 3.2. *For $n \geq 1$, the following identity holds*

$$2(2n+1)b(n+1) - nb(n) = \left(7n^2 + 8n + 2\right)C_n^2. \tag{40}$$

Proof. We prove the identity by the induction method. For $n = 1$, we directly check the identity. Suppose that the identity holds for a given number n. Since $(n+2)C_{n+1} = 2(2n+1)C_n$, we have that

$$\begin{aligned}
&\left(7n^2 + 8n + 2\right)\left(7n^2 + 22n + 17\right)(n+2)^2 C_{n+1}^2 \\
&= \left(7n^2 + 8n + 2\right)\left(7n^2 + 22n + 17\right)4(2n+1)^2 C_n^2 \\
&= 4(2n+1)^2\left(7n^2 + 22n + 17\right)(2(2n+1)b(n+1) - nb(n)) \\
&= 8(2n+1)^3\left(7n^2 + 22n + 17\right)b(n+1) + q_3(n+2)b(n) \\
&= q_1(n+2)b(n+2) + \left(8(2n+1)^3\left(7n^2 + 22n + 17\right) - q_2(n+2)\right)b(n+1) \\
&= 2\left(7n^2 + 8n + 2\right)(2n+3)(n+2)^2 b(n+2) - \left(7n^2 + 8n + 2\right)(n+2)^2(n+1)b(n),
\end{aligned}$$

where we have applied the recurrence relation (36), we obtain the identity for $n+1$, and we conclude the result. □

Sequences $(a(n))_{n \geq 0}$ and $(b(n))_{n \geq 1}$ are jointly connected as the next lemma shows. The proof is left to the reader.

Lemma 3.3. *For $n \geq 1$, the following two identities hold*

$$\begin{vmatrix} q_1(n) & q_3(n) \\ p_1(n-1) & p_3(n-1) \end{vmatrix} = -8Q(n)(2n-1)(2n-3)^2(n-2); \tag{41}$$

$$\begin{vmatrix} q_1(n) & q_2(n) \\ p_1(n-1) & p_2(n-1) \end{vmatrix} = 16Q(n)(2n-1)(2n-3)^3, \tag{42}$$

where $Q(n) := 147n^4 - 546n^3 + 666n^2 - 293n + 34$.

Our last aim of this section is to show an alternative of the following identity

$$\binom{2n}{n}^2 = \sum_{k=0}^{n} \frac{3n-2k}{n} \binom{2n-1-k}{n-1}^2, \tag{43}$$

in Theorem 3.4. An original proof is presented in ([22], Theorem 2.3 (ii)), and it is a straightforward consequence of a more general identity in combinatorial numbers ([22], Theorem 2.3 (i)). The proof which we present here allows to recognize the natural connection among the sequences $(a(n))_{n\geq 0}$ and $(b(n))_{n\geq 1}$ and the Catalan numbers $(C_n)_{n\geq 0}$. Note that one may rewrite the identity (43) in an equivalent way.

Theorem 3.4. *For* $n \geq 1$, *the following identity holds*

$$((n+1)C_n)^2 = 4(3a(n-1) - 2b(n)), \qquad n \geq 1. \tag{44}$$

Proof. We write by $c(n) = ((n+1)C_n)^2 = \binom{2n}{n}^2$, and then we have to check the following identity

$$c(n) = 4(3a(n-1) - 2b(n)), \qquad n \geq 1, \tag{45}$$

where sequences $(a(n))_{n\geq 0}$ and $(b(n))_{n\geq 1}$ are considered in the second section. Note that

$$\begin{aligned}
&p_1(n-1)q_1(n)4(3a(n-1) - 2b(n)) \\
&= 12q_1(n)(p_2(n-1)a(n-2) + p_3(n-1)a(n-3)) \\
&\quad -8p_1(n-1)(q_2(n)b(n-1) + q_3(n)b(n-2)) \\
&= 12a(n-2)(p_1(n-1)q_2(n) + 16Q(n)\left(2(n-1)(2n-3)^3\right) \\
&\quad +12a(n-3)(p_1(n-1)q_3(n) - 8Q(n)(2n-1)(2n-3)(n-2)) \\
&\quad -8p_1(n-1)(q_2(n)b(n-1) - 2p_1(n-1)q_3(n)b(n-2) \\
&= p_1(n-1)q_2(n)(12a(n-2) - 8b(n-1)) \\
&\quad +p_1(n-1)q_3(n)(12a(n-3) - 8b(n-2)) \\
&\quad +96(2n-1)(2n-3)^2Q(n)(2(2n-3)a(n-2) - (n-2)a(n-3)),
\end{aligned}$$

where we have applied the recurrence relations (30) and (36) and Lemma 3.3. By the induction method and Theorem 3.1, we have that

$$\begin{aligned}
p_1(n-1)q_1(n)4(3a(n-1) - 2b(n)) &= p_1(n-1)q_2(n)c(n-1) + p_1(n-1)q_3(n)c(n-2) \\
&\quad +24(2n-1)(2n-3)^2Q(n)(21n-34)c(n-2)
\end{aligned}$$

for $n \geq 2$. Since $4(2n-3)^2c(n-2) = (n-1)^2c(n-1)$ for $n \geq 2$, we have that

$$24(2n-1)(2n-3)^2Q(n)(21n-34)c(n-2) = 3p_1(n-1)Q(n)c(n-1), \quad n \geq 1. \tag{46}$$

Finally, we get that

$$\begin{aligned}
&p_1(n-1)q_1(n)4(3a(n-1) - 2b(n)) \\
&\qquad = p_1(n-1)c(n-1)\left(q_2(n) + q_3(n)\frac{(n-1)^2}{4(2n-3)^2} + 3Q(n)\right)
\end{aligned}$$

and

$$c(n-1)\left(q_2(n)+q_3(n)\frac{(n-1)^2}{4(2n-3)^2}+3Q(n)\right)$$
$$=c(n-1)8\left(7n^2-20n+14\right)(2n-1)^3$$
$$=c(n)n^2 2\left(7n^2-20n+14\right)(2n-1)=c(n)q_1(n),$$

and we conclude the proof. □

4. Moments of squares and cubes of Catalan triangle numbers

In this section, we present some moments of squares and cubes of Catalan triangle numbers $(B_{n,k})_{n\geq 1, n\geq k\geq 1}$ and $(A_{n,k})_{n\geq 1, n+1\geq k\geq 1}$, i.e.,

$$\sum_{k=1}^{n} k^m B_{n,k}^j, \qquad \sum_{k=1}^{n+1} (2k-1)^m A_{n,k}^j, \tag{47}$$

for $j=2,3$ and $m\in\mathbb{N}$. For $m=0$, these identities are shown in [14, 24]. See a unified proof in ([22], Corollary 2.2).

Proposition 4.1. *For* $n\geq 1$*, we have*

i. $\sum_{k=1}^{n} B_{n,k}^2 = C_{2n-1}$,

ii. $\sum_{k=1}^{n+1} A_{n,k}^2 = C_{2n}$,

iii. $\sum_{k=1}^{n} (-1)^k B_{n,k}^2 = -\frac{n+1}{2} C_n$.

Remark. The first values of $\sum_{k=1}^{n+1}(-1)^k A_{n,k}^2$ are

$$0, \quad 4, \quad 32, \quad 236, \quad 1865, \quad 16080, \tag{48}$$

for $1\leq n\leq 6$. We are not able to find any closed formula for the general expression.

In ([13], Theorem 2), the closed expression of

$$\Omega_m(n) := \sum_{k=1}^{n} k^m B_{n,k}^2, \tag{49}$$

is given for $m\in\mathbb{N}\cup\{0\}$. We present now for $m\in\{0,1,\cdots,7\}$. Previously, the WZ theory was used to show them in ([12], Theorem 2.1, 2.2). See also ([1], Section 5).

Theorem 4.2. *For* $n\in\mathbb{N}$,

i. $\Omega_0(n) = C_{2n-1}$

$$\Omega_2(n) = \frac{(3n-2)n}{4n-3}C_{2n-1},$$

$$\Omega_4(n) = \frac{(15n^3-30n^2+16n-2)n}{(4n-3)(4n-5)}C_{2n-1},$$

$$\Omega_6(n) = \frac{(105n^5-420n^4+588n^3-356n^2+96n-10)n}{(4n-3)(4n-5)(4n-7)}C_{2n-1}.$$

ii. $\Omega_1(n) = (2n-3)(n+1)C_nC_{n-2},$

$$\Omega_3(n) = n(2n-3)(n+1)C_nC_{n-2},$$

$$\Omega_5(n) = n\left(3n^2-5n+1\right)(n+1)C_nC_{n-2},$$

$$\Omega_7(n) = n\left(6n(n-1)^2-1\right)(n+1)C_nC_{n-2}.$$

In ([13], Theorem 4, 8), the closed expression of

$$\Psi_m(n) := \sum_{k=1}^{n}(2k-1)^m A_{n,k}^2, \tag{50}$$

is obtained for $m \in \mathbb{N} \cup \{0\}$. Now, we present the particular cases for $m \in \{0, 1, \cdots, 7\}$ in the next theorem.

Theorem 4.3. *For* $n \in \mathbb{N}$,

i. $\Psi_0(n) = C_{2n},$

$$\Psi_2(n) = \frac{-1+4n+12n^2}{4n-1}C_{2n},$$

$$\Psi_4(n) = \frac{3-16n-104n^2+240n^4}{(4n-1)(4n-3)}C_{2n},$$

$$\Psi_6(n) = \frac{-15+92n+1116n^2+2080n^3-4368n^4-6720n^5+6720n^6}{(4n-1)(4n-3)(4n-5)}C_{2n}.$$

ii. $\Psi_1(n) = (n+1)C_nC_{n-1}(4n-2),$

$$\Psi_3(n) = (n+1)C_nC_{n-1}\left(16n^2-2\right),$$

$$\Psi_5(n) = (n+1)C_nC_{n-1}\left(96n^3+32n^2-4n-2\right),$$

$$\Psi_7(n) = (n+1)C_nC_{n-1}\frac{1536n^5-1536n^4-960n^3-160n^2+20n+6}{2n-3}.$$

Integer sequences of numbers $(a(n))_{n\geq 0}$ and $(b(n))_{n\geq 1}$ were treated in Section 3. They play a very interesting role to describe the sums of cubes of Catalan triangle numbers, as the next result shows. See proofs and more details in ([22], Section 3).

Theorem 4.4. *For* $n \geq 1$, *we have*

i. $\sum_{k=0}^{n} B_{n,k}^3 = \frac{n+1}{2}C_nb(n),$

ii. $\sum_{k=1}^{n+1} A_{n,k}^3 = (n+1)C_n\left(\left(2(n+1)C_n\right)^2 - 3a(n)\right),$

iii. $\sum_{k=1}^{n+1}(-1)^k A_{n,k}^3 = \frac{n-1}{2n+1}\binom{2n}{n}\binom{3n}{n}$.

Remark. To check $\sum_{k=1}^{n} B_{n,k}^3$ in Theorem 4.4 (i), we need to show the identity:

$$\binom{2n}{n}^2 = \sum_{k=0}^{n} \frac{3n-2k}{n}\binom{2n-1-k}{n-1}^2, \qquad n \geq 1, \tag{51}$$

see ([22], Theorem 3.3). In Theorem 3.4, we have presented an alternative proof of this identity.

The first values of $\sum_{k=1}^{n}(-1)^k B_{n,k}^3$ are

$$-1, \quad -7, \quad -62, \quad -215, \quad 17332, \quad 945342, \tag{52}$$

for $1 \leq n \leq 6$. We are not able to find any closed formula for the general expression.

5. Conclusions and future developments

In this paper we have studied in detail

$$\sum_{k=1}^{n} k^m B_{n,k}^j, \qquad \sum_{k=1}^{n+1}(2k-1)^m A_{n,k}^j, \tag{53}$$

for $n \in \mathbb{N}$ and several values of $j \in \mathbb{N}$. The main objective is to give a closed formula where a factor is $\frac{n+1}{2}C_n$, $(n+1)C_n$, C_{2n}, or other Catalan number, for example, in Theorem 2.1, Proposition 4.1, and Theorems 4.2 and 4.3. These results complete previous studies for $m = 0, 1$ and 2. In the case of $j = 3$ and $m = 0$, some known integer sequences $(a(n))_{n \geq 0}$ and $(b(n))_{n \geq 1}$ appear in Theorem 4.4. Also the alternating sums

$$\sum_{k=1}^{n}(-1)^k B_{n,k}^j, \qquad \sum_{k=1}^{n+1}(-1)^k A_{n,k}^j, \tag{54}$$

are considered in Theorem 2.2, Proposition 4.1 (iii), and Proposition 4.4 (iii).

To show these identities, we have combined the analytic proofs and the WZ theory which is useful to show combinatorial identities. Our results allow continuing this research, and future developments could be made.

In the following, we present some conjectures about new identities in Catalan triangle numbers. These conjectures are about the properties of divisibility of sums and alternating sums of powers of Catalan triangle numbers $B_{n,k}$ and $A_{n,k}$. The factors which we consider are $\frac{n+1}{2}C_n$ and $(n+1)C_n$.

Conjecture 5.1. After Theorem 2.1 (i) and (ii), it is natural to conjecture that for $m, n \in \mathbb{N}$

$$\Delta_{2m}(n) = \frac{n+1}{2}C_n n P_{m-1}(n), \tag{55}$$

$$\Delta_{2m-1}(n) = 2^{n-m-1}Q_{m-1}(n), \tag{56}$$

where P_{m-1} and Q_{m-1} are polynomials of integer coefficients at degree at most $m-1$.

Conjecture 5.2. After Theorem 2.1 (iii) and (iv), it is also natural to conjecture that for $m, n \in \mathbb{N}$,

$$\Lambda_{2m}(n) = (n+1)C_n R_{m-1}(n), \tag{57}$$

$$\Lambda_{2m-1}(n) = 2^{2n} S_{m-1}(n), \tag{58}$$

where R_{m-1} and S_{m-1} are polynomials of integer coefficients at degree at most $m-1$.

Conjecture 5.3. In **Table 1**, we present the moments $\sum_{k=1}^{n} k^m B_{n,k}^3$ for $m \in \{1,2,3,4\}$ and $n \in \{1,2,3,4,5\}$. Then we conjecture that the factor $\frac{n+1}{2} C_n$ divides $\sum_{k=0}^{n} k^{2m} B_{n,k}^3$ for $m, n \in \mathbb{N}$.

Conjecture 5.4. In **Table 2**, we give the moments $\sum_{k=1}^{n+1} (2k-1)^m A_{n,k}^3$ for $m \in \{1,2,3,4\}$ and $n \in \{1,2,3,4,5\}$. We conjecture that the factor $(n+1)C_n$ divides $\sum_{k=0}^{n+1} (2k-1)^{2m} A_{n,k}^3$ for $m, n \in \mathbb{N}$.

Conjecture 5.5. We give the moments $\sum_{k=1}^{n} k^m B_{n,k}^4$ for $m \in \{1,2,3,4\}$ and $n \in \{1,2,3,4,5\}$ in **Table 3**. Then we conjecture that the factor $\frac{n+1}{2} C_n$ divides $\sum_{k=0}^{n} k^{2m-1} B_{n,k}^4$ for $m, n \in \mathbb{N}$.

Conjecture 5.6. In **Table 4**, we give the moments $\sum_{k=0}^{n+1} (2k-1)^m A_{n,k}^4$ for $m \in \{1,2,3,4\}$ and $n \in \{1,2,3,4,5\}$. We conjecture that $(n+1)C_n$ divides $\sum_{k=0}^{n+1} (2k-1)^{2m-1} A_{n,k}^4$ for $m, n \in \mathbb{N}$.

Conjecture 5.7. The sums of alternating powers of Catalan triangle numbers $B_{n,k}$ and $A_{n,k}$,

n	$\sum_{k=1}^{n} kB_{n,k}^3$	$\sum_{k=1}^{n} k^2 B_{n,k}^3$	$\sum_{k=1}^{n} k^3 B_{n,k}^3$	$\sum_{k=1}^{n} k^4 B_{n,k}^3$
1	1	1	1	1
2	10	12	16	24
3	256	390	664	1230
4	8884	15, 680	30, 592	64, 400
5	356, 374	701, 820	1, 523, 158	3, 569, 580

Table 1.
Moments of cubes of $B_{n,k}$.

n	$\sum_{k=1}^{n+1} (2k-1) A_{n,k}^3$	$\sum_{k=1}^{n+1} (2k-1)^2 A_{n,k}^3$	$\sum_{k=1}^{n+1} (2k-1)^3 A_{n,k}^3$	$\sum_{k=1}^{n+1} (2k-1)^4 A_{n,k}^3$
1	4	10	28	82
2	94	276	862	2820
3	2944	9860	35, 776	139, 700
4	111, 010	417, 200	1, 713, 826	7, 610, 960
5	4, 677, 160	19, 342, 008	87, 730, 360	430, 535, 448

Table 2.
Moments of cubes of $A_{n,k}$.

$$\sum_{k=1}^{n}(-1)^k B_{n,k}^j, \qquad \text{and} \qquad \sum_{k=1}^{n+1}(-1)^k A_{n,k}^j, \tag{59}$$

have been considered in this paper: in Theorem 2.2 (i) and (ii) for $j = 1$, in Proposition 4.1 (iii) for $j = 2$, and in Theorem 4.4 (iii) for $j = 3$. In **Table 5**, we present the alternating sums of the fourth and fifth powers of Catalan triangle numbers. All these results join to conjecture that the factor $\frac{n+1}{2}C_n$ divides $\sum_{k=0}^{n}(-1)^k B_{n,k}^{2m}$ for $m, n \in \mathbb{N}$ and $(n+1)C_n$ divides $\sum_{k=0}^{n+1}(-1)^k A_{n,k}^{2m-1}$ for $m, n \in \mathbb{N}$.

Finally we give some general comments and ideas which could be followed in future works.

i. The generating formula (1) allows an interesting way to show some combinatorial identities in an analytic way.

n	$\sum_{k=1}^{n} kB_{n,k}^4$	$\sum_{k=1}^{n} k^2B_{n,k}^4$	$\sum_{k=1}^{n} k^3B_{n,k}^4$	$\sum_{k=1}^{n} k^4B_{n,k}^4$
1	1	1	1	1
2	18	20	24	32
3	1140	1658	2700	4802
4	119, 140	203, 760	380, 800	758, 304
5	15, 339, 240	29, 193, 890	60, 190, 200	132, 142, 274

Table 3.
Moments of the fourth power of $B_{n,k}$.

n	$\sum_{k=1}^{n+1}(2k-1)A_{n,k}^4$	$\sum_{k=1}^{n+1}(2k-1)^2A_{n,k}^4$	$\sum_{k=1}^{n+1}(2k-1)^3A_{n,k}^4$	$\sum_{k=1}^{n+1}(2k-1)^4A_{n,k}^4$
1	4	10	28	82
2	264	770	2328	7202
3	23, 440	75, 348	256, 240	925, 092
4	2, 699, 200	9, 688, 050	37, 458, 400	155, 596, 914
5	368, 708, 256	1, 458, 679, 508	6, 249, 158, 496	28, 738, 974, 308

Table 4.
Moments of the fourth power of $A_{n,k}$.

n	$\sum_{k=1}^{n}(-1)^kB_{n,k}^4$	$\sum_{k=1}^{n+1}(-1)^kA_{n,k}^4$	$\sum_{k=1}^{n}(-1)^kB_{n,k}^5$	$\sum_{k=1}^{n+1}(-1)^kA_{n,k}^5$
1	−1	0	−1	0
2	−15	64	−31	210
3	−370	5312	−2102	52, 800
4	−1295	418, 640	−7775	13, 489, 350
5	1, 669, 374	32, 351, 744	109, 796, 596	3, 453, 624, 720

Table 5.
Sums of alternating powers of $B_{n,k}$ and $A_{n,k}$.

ii. Alternating moments of Catalan triangle numbers $B_{n,k}$ and $A_{n,k}$, i.e.,

$$\sum_{k=1}^{n}(-k)^m B_{n,k}^j, \qquad \sum_{k=1}^{n+1}(-(2k-1))^m A_{n,k}^j, \tag{60}$$

are a new interesting research which could be considered in later articles, compared with ([24], Theorem 2.3).

iii. In a similar way, weight moments of Catalan triangle numbers $B_{n,k}$ and $A_{n,k}$,

$$\sum_{k=1}^{n} a^k B_{n,k}^j, \qquad \sum_{k=1}^{n+1} b^k A_{n,k}^j, \qquad j, n \in \mathbb{N}, \tag{61}$$

are worth studying them for some $a, b \in \mathbb{N}$, compared with ([9], Theorem 1.1).

Acknowledgements

P.J. Miana has been partially supported by Project MTM2016-77710-P, DGI-FEDER, of the MCYTS and Project E26-17R, D.G. Aragón, Spain. Natalia Romero has been partially supported by the Spanish Ministry of Science, Innovation and Universities, Project PGC2018-095896-B-C21.

Appendix

In this appendix, we present some tables of powers of Catalan triangle numbers $B_{n,k}$ and $A_{n,k}$. As we have mentioned above, they are used to conjecture some statements in the Section 5.

Additional information

Mathematics Subject Classification: 05A19; 05A10; 11B65, 11B75

Author details

Pedro J. Miana[1]* and Natalia Romero[2]

1 Departamento de Matemáticas, Instituto Universitario de Matemáticas y Aplicaciones, Universidad de Zaragoza, Zaragoza, Spain

2 Departamento de Matemáticas y Computación, Universidad de La Rioja, Logroño, Spain

*Address all correspondence to: pjmiana@unizar.es

References

[1] Chen X, Chu W. Moments on Catalan numbers. Journal of Mathematical Analysis and Applications. 2009;**349**(2):311-316

[2] Chu W. A new combinatorial interpretation for generalized Catalan numbers. Discrete Mathematics. 1987; **65**:91-94

[3] Chu W. Summation formulae involving harmonic numbers. Filomat. 2012;**26**(1):143-152

[4] Deutsch E, Shapiro L. A survey of the Fine numbers. Discrete Mathematics. 2001;**241**:241-265

[5] Eplett WJR. A note about the Catalan triangle. Discrete Mathematics. 1979;**25**: 289-291

[6] Guo VJW, Zeng J. Factors of binomial sums from Catalan triangle. Journal of Number Theory. 2010;**130**(1): 172-186

[7] Guo VJW, Lian X. Proofs of two conjectures on Catalan triangle numbers. Journal of Difference Equations and Applications. 2018;**24**(9): 1473-1487

[8] Gutiérrez JM, Hernández MA, Miana PJ, Romero N. New identities in the Catalan triangle. Journal of Mathematical Analysis and Applications. 2008;**341**(1):52-61

[9] Lee K-H, Oh S-J. Catalan triangle numbers and binomial coefficients. Contemporary Mathematics. 2018;**713**

[10] Lang W. On polynomials related to powers of the generating function of Catalans numbers. Fibonacci Quarterly. 2000;**38**(5):408-419

[11] Hilton P, Pedersen J. Catalan numbers, their generalization and their uses. Mathematical Intelligencer. 1991; **13**:64-75

[12] Miana PJ, Romero N. Computer proofs of new identities in the Catalan triangle. In: Biblioteca de la Revista Matemática Iberoamericana. Proc. of the "Segundas Jornadas de Teoría de Números"; Madrid. 2007. pp. 203-208

[13] Miana PJ, Romero N. Moments of combinatorial and Catalan numbers. Journal of Number Theory. 2010; **130**(8):1876-1887

[14] Shapiro LW. A Catalan triangle. Discrete Mathematics. 1976;**14**:83-90

[15] Slavík A. Identities with squares of binomial coefficients. Ars Combinatoria. 2014;**113**:377-383

[16] Sloane N. Available from: http:// www.research.att.com/

[17] Sloane N. A Handbook of Integer Sequences. New York: Academic Press; 1973

[18] Spies J. Some identities involving harmonic numbers. Mathematics of Computation. 1990;**55**:839-863

[19] Stanley RP. Enumerative Combinatorics. Vol. 2. Cambridge: Cambridge University Press; 1999

[20] Stanley RP. Catalan Numbers. Cambridge: Cambridge University Press; 2015

[21] Aigner M. Catalan-like numbers and determinants. Journal of Combinatorial Theory, Series A. 1999;**87**:33-51

[22] Miana PJ, Ohtsuka H, Romero N. Sums of powers of Catalan triangle numbers. Discrete Mathematics. 2017; **340**(10):2388-2397

[23] Brent RP, Ohtsuka H, Osborn J-AH, Prodinger H. Some binomial sums involving absolute values. Journal of

Integer Sequences. 2016;**19**; Article 16.3.7

[24] Zhang Z, Pang B. Several identities in the Catalan triangle. Indian Journal of Pure and Applied Mathematics. 2010; **41**(2):363-378

[25] Petkovsek M, Wilf HS, Zeilberger D. $A = B$. Wellesley: A. K. Peters Ltd.; 1997. Available from: http://www.cis.upenn.edu/~wilf/AeqB.html

[26] Paule P, Schneider C. Computer proofs of a new family of harmonic number identities. Advances in Applied Mathematics. 2003;**31**(2):359-378

[27] Wilf H, Zeilberger D. Rational functions certify combinatorial. Journal of the American Mathematical Society. 1990;**3**:147-158

[28] Benjamin AT, Quinn J. Proofs that Really Count, Dolciani Mathematical Exposition 27. Washington, DC: Mathematical Association of America; 2003

Section 2

Applications

Chapter 6

Modular Sumset Labelling of Graphs

Sudev Naduvath

Abstract

Graph labelling is an assignment of labels or weights to the vertices and/or edges of a graph. For a ground set X of integers, a sumset labelling of a graph is an injective map $f : V(G) \to \mathcal{P}(X)$ such that the induced function $f^{\oplus} : E(G) \to \mathcal{P}(X)$ is defined by $f^{+}(uv) = f(u) + f(v)$, for all $uv \in E(G)$, where $f(u) + f(v)$ is the sumset of the set-label, the vertices u and v. In this chapter, we discuss a special type of sumset labelling of a graph, called modular sumset labelling and its variations. We also discuss some interesting characteristics and structural properties of the graphs which admit these new types of graph labellings.

Keywords: sumset labelling, modular sumset labelling, weak modular sumset labelling, strong modular sumset labelling, arithmetic modular sumset labelling

1. Introduction

For terminology and results in graph theory, we refer to [1–5]. For further notions and concepts on graph classes, graph operations, graph products and derived graphs, refer to [3, 6–10]. Unless mentioned otherwise, all graphs mentioned in this chapter are simple, finite, connected and undirected.

1.1 Basics of graph labelling

Labelling of a graph G can broadly be considered as an assignment of labels or weights to the elements (vertices and edges) of G subject to certain pre-defined conditions. The research on graph labelling has flourished in the second half of twentieth century after the introduction of the notion of β-valuations of graphs in [11]. The *β-valuation* of a graph G is an injective map $f : V(G) \to \{1, 2, 3, \ldots, |E|\}$ such that the induced function $f^*: E(G) \to \{1, 2, 3, \ldots, |E|\}$, defined by $f^*(uv) = |f(u) - f(v)|$ for all $uv \in E(G)$, is also injective. Later, β-valuation of graphs was popularly known to be the *graceful labelling* of graphs (see [12]). Many variations of number valuations have been defined in the literature since then and most of those studies were based on the number theory and/or number theoretic properties of sets. For concepts and results in number theory, see [13–16].

Analogous to the number valuations of graphs, the notion of set-labelling of graphs has been introduced in [17] as follows: Given a non-empty ground set X, a *set-labelling* or a *set-valuation* of a graph G is an injective function $f : V(G) \to \mathcal{P}(X)$, the power set of X, such that the induced function $f^{\oplus} : V(G) \to \mathcal{P}(X)$ defined by

$f^{\oplus}(uv) = f(u) \oplus f(v)$, for all $uv \in E(G)$, where $\bigoplus$ is the symmetric difference of two sets. A graph which admits a set-labelling is called a *set-labelled graph* or a *set-valued graph*. If the induced function $f^{\oplus}$ is also injective, then the set-labelling f is called a *set-indexer*.

Subsequent to this study, many intensive investigations on set-labelling of graphs and its different types have been taken place. An overview of such studies on set-labelling of graphs can be seen in [17–23]. The study on set-labelling has been extended by replacing the binary operation $\bigoplus$ by some other binary operations of sets. For example, two different types of set-labellings—called disjunctive set-labelling and conjunctive set-labelling—of graphs have been studied in [24]. These set-labellings are defined respectively in terms of the union and the intersection of two sets instead of the symmetric difference of two sets.

1.2 Sumsets and integer additive set-labelled graphs

Integer additive set-labelling or sumset labelling of graphs has been a new addition to the theory of set-labelling of graphs recently. The notion of sumsets of two sets is explained as follows: Let A and B be two sets of numbers. The *sumset* of A and B is denoted by $A + B$ and is defined by $A + B = \{a + b : a \in A, b \in B\}$ (see [25]). Remember that a sumset of two sets can be determined if and only if both of them are number sets. If either A or B is countably infinite, then their sumset $A + B$ is also a countably infinite set and if any one of them is a null set, then the sumset is also a null set. If C is the sumset of two sets A and B, then both A and B are said to be the *summands* of C.

We note that $A + \{0\} = A$ and hence A and $\{0\}$ are called the trivial summands of the set A. Also, note that $A + B$ need not be a subset or a super set of A and/or B. But, $A \subset A + B$ if $0 \in B$. Furthermore, the sumset of two subsets of a set X need not be a subset of the ground set X. These observations are clear deviations from the other common binary operations of sets and thus the study of sumsets becomes more interesting. For the terms, concepts and results on sumsets, we refer to [25–34].

Note that if A and B are two non-empty finite sets of integers, then $|A| + |B| - 1 \leq |A + B| \leq |A \times B| = |A||B|$ (see [25]). The exact cardinality of the sumset $A + B$ always depends on the number as well as the pattern of elements in both the summands A and B. The counting procedure in this case is explained in [35] as follows: Two ordered pairs (a, b) and (c, d) in $A \times B$ is said to be *compatible* if $a + b = c + d$. If (a, b) and (c, d) are compatible, then it is written as $(a, b) \sim (c, d)$. It can easily be verified that this relation is an equivalence relation. A *compatibility class* of an ordered pair (a, b) in $A \times B$ with respect to the integer $k = a + b$ is the subset of $A \times B$ defined by $\{(c, d) \in A \times B : (a, b) \sim (c, d)\}$ and is denoted by $[(a, b)]_k$ or C_k. The cardinality of a compatibility class in $A \times B$ lies between 1 and $\min\{|A|, |B|\}$. Note that the sum of coordinates of all elements in a compatibility class is the same and this sum will be an element of the sumset $A + B$. That is, the cardinality of the sumset of two sets is equal to the number of equivalence classes on the Cartesian product of the two sets generated by the compatibility relation defined on it.

Using the concepts of the sumsets of sets, the notion of integer additive set-labelling of graphs has been introduced in [36] as follows: Let X be a set of non-negative integers and $\mathcal{P}_0(X)$ be the collection of the non-empty subsets of X. Then, an *integer additive set-labelling* or an *integer additive set-valuation* of a graph G is an injective map $f : V(G) \to \mathcal{P}(X)$ such that the induced function $f^{\oplus} : E(G) \to \mathcal{P}(X)$ is defined by $f^{+}(uv) = f(u) + f(v)$, for all $uv \in E(G)$, where $f(u) + f(v)$ is the sumset of the set-label the vertices u and v (see [36, 37]). A graph with an integer additive set-labelling is called an *integer additive set-labelled graph*. It can very easily be verified that every graph G admits an integer additive set-labelling, provided the ground set X is chosen judiciously.

Following to the above path-breaking study, the structural properties and characteristics of different types of integer additive set-labellings of graphs are studied intensively in accordance with the cardinality of the set-label, nature and pattern of elements in the set-label, nature of the collection of set-label, etc. Some interesting and significant studies in this area can be found in [35–44]. Later, the studies in this area have been extended by including the sets of integers (including negative integers also) for labelling the elements of a graph. Some extensive studies in this area, can be seen in [45, 46].

As a specialisation of the sumset labelling of graphs, the notion of modular sumset labelling of graphs and corresponding results are discussed in the following section.

2. Modular sumset labelling of graphs

2.1 Basics of modular sumsets

Recall that $\mathbb{Z}_n$ denotes the set of integers modulo n, where n is a positive integer. The *modular sumset* of two subsets A and B of $\mathbb{Z}_n$ is the set $\{k : a \in A, b \in B, a + b \equiv k \pmod{n}\}$. Unless mentioned otherwise, throughout this chapter, the notation $A + B$ denotes the modular sumset of the sets A and B. Unlike the ordinary sumsets, the modular sumset $A + B \subseteq \mathbb{Z}_n$ if and only if $A, B \subseteq \mathbb{Z}_n$. This fact will ease many restrictions imposed on the vertex set-label of a sumset graph G in order to ensure that the edge set-label are also subsets of the ground set.

If we assign the null set Ø to any vertex as the set-label, the set-label of every edge incident at that vertex will also be a null set. To avoid such an embarrassing situation, we do not consider the null set for labelling any vertex of graphs. Thus, the set of all non-empty subsets of a set X is denoted by $\mathcal{P}_0(X)$. That is, $\mathcal{P}_0(X) = \mathcal{P}(X) \backslash \{0\}$.

2.2 Modular sumset graphs

In view of the facts stated above, the modular sumset labelling of a graph is defined as follows:

Definition 2.1. [47] A modular sumset labelling of a graph G is an injective function $f : V(G) \to \mathcal{P}_0(\mathbb{Z}_n)$ such that the induced function $f^+ : E(G) \to \mathcal{P}_0(\mathbb{Z}_n)$ is defined as $f^+(uv) = f(u) + f(v)$, where $f(u) + f(v)$ is the modular sumset of the set labels of the vertices u and v. A graph which admits a modular sumset labelling is called a *modular sumset graph*.

Definition 2.2. [47] A modular sumset labelling of a graph G is said to be a *uniform modular sumset labelling* of G if the set-label of all its edges have the same cardinality. A modular sumset labelling f of G is said to be a *k-uniform modular sumset labelling* if $f^+(uv) = k, \forall uv \in E(G)$.

Proposition 2.1. [47] *Every graph G admits modular sumset labelling (for a suitable choice of n).*

The proof of the above proposition is immediate from the fact that $f(u) + f(v) \subseteq \mathbb{Z}_n$ if and only if $f(u), f(v) \subseteq \mathbb{Z}_n$.

An immediate question that arises in this context is about the minimum size of the ground set $\mathbb{Z}_n$ (that is, the minimum value of n) required for the existence of a modular sumset labelling of G.

As in the case of sumsets, the cardinality of the modular sumsets also attracted the attention. Hence, we have the bounds for the cardinality of an edge set-label of a modular sumset graph G is as follows:

Theorem 2.2. [47] *Let $f : V(G) \to \mathcal{P}_0(\mathbb{Z}_n)$ be a modular sumset labelling of a given graph G. Then, for any edge $uv \in E(G)$, we have*

$$|f(u)| + |f(v)| - 1 \leq |f^+(uv)| = |f(u) + f(v)| \leq |f(u)|\,|f(v)| \leq n. \tag{1}$$

The theorem follows immediately from the theorem on the cardinality of sumsets (see Theorem 2.7, p. 52, [25]).

In this context, it is quite interesting to investigate whether the bounds are sharp. It has also been proved in [25] that the lower bound is sharp when both $f(u)$ and $f(v)$ are arithmetic progressions (we call set an arithmetic progression if its elements are in arithmetic progression) with the same common difference. We shall discuss the different types of modular sumset graphs based on the set-labelling numbers of its vertices and edges, one by one in the coming discussions.

3. Arithmetic modular sumset graphs

As mentioned above, the lower bound of the inequality (1) is sharp if both summand set-label are arithmetic progressions with the same common difference. If the context is clear, the common difference of the set-label (if exists) of an element may be called the *common difference* of that element. The *deterministic ratio* of an edge of G is the ratio, $k \geq 1$ between the common differences of its end vertices. In view of this terminology we have the following definition.

Definition 3.1. For any vertex v of G, if $f(v)$ is an arithmetic progression, then the modular sumset labelling $f : V(G) \to \mathcal{P}_0(\mathbb{Z}_n)$ is called a *vertex arithmetic modular sumset labelling* of G. In a similar manner, for any edge e of G, if $f(e)$ is an arithmetic progression, then the modular sumset labelling $f : E(G) \to \mathcal{P}_0(\mathbb{Z}_n)$ is called an *edge arithmetic modular sumset labelling* of G.

The difference set of a non-empty set A, denoted by D_A, is the set defined by $D_A = \{|a - b| : a, b \in A\}$. Note that if A is an arithmetic progression, then its difference set D_A is also an arithmetic progression and vice versa. Analogous to the corresponding result of the edge-arithmetic sumset labelling of graphs (see [44, 46]), the following result is a necessary and sufficient condition for a graph G to be edge-arithmetic modular sumset graph in terms of the difference sets the set-label of vertices of G.

Theorem 3.1. *Let f be a modular sumset labelling defined on a graph G. If the set-label of an edge of G is an arithmetic progression if and only if the sumset of the difference sets of set-label of its end vertices is an arithmetic progression.*

Proof. Let $f : V(G) \to \mathcal{P}(\mathbb{N}_0)$ be a modular sumset labelling defined on G. Let a_i, a_j be two arbitrary elements in $f(u)$ and let b_r, b_s be two elements in $f(v)$. Then, $|a_i - a_j| \in D_{f(u)}$ and $|a_i - a_j| \in D_{f(u)}$. That is, $D_{f(u)} = \{|a_i - a_j| : a_i, a_j \in f(u)\}$ and $D_{f(v)} = \{|b_r - b_s| : b_r, b_s \in f(v)\}$.

Now, assume that $f^+(e) = f^+(uv)$ is an arithmetic progression for an edge $e = uv \in E(G)$. That is, $A = f(u) + f(v)$ is an arithmetic progression. Then, the difference set $D_A = \{|a - b| : a, b \in A = f(u) + f(v)\}$ is also an arithmetic progression. Since $a, b \in A$, we have $a = a_i + b_r$ and $b = a_j + b_s$, where $a_i, a_j \in f(u)$ and $b_r, b_s \in f(v)$. Then,

$$\begin{aligned}D_A &= \{|a-b|: a,b \in A\}\\ &= \{|a_i + b_r - (a_j + b_s)|: a_i, a_j \in f(u), b_r, b_s \in f(v)\}\\ &= \{|a_i - a_j| + |b_r - b_s|: a_i, a_j \in f(u), b_r, b_s \in f(v)\}\\ &= \{|a_i - a_j|: a_i, a_j \in f(u)\} + \{|b_r - b_s|: b_r, b_s \in f(v)\}\\ &= D_{f(u)} + D_{f(v)}\end{aligned}$$

Hence, $D_{f(u)} + D_{f(v)}$ is an arithmetic progression.

Conversely, assume that $D_{f(u)} + D_{f(v)}$ is an arithmetic progression. Then, by previous step, we have $D_{f(u)} + D_{f(v)} = D_A$, where $A = f(u) + f(v)$. Then, we have D_A is an arithmetic progression. Since the difference set D_A is an arithmetic progression, then by the above remark, we have $A = f(u) + f(v) = f^+(uv)$ is also an arithmetic progression. Hence, the edge $e = uv$ has an arithmetic progression as its set-label. □

In view of the notions mentioned above, we note that there are some graphs, all whose elements have arithmetic progressions as their set-label and there are some graphs, the set-label of whose edges are not arithmetic progressions. Keeping this in mind, we define the following notion.

Definition 3.2. An *arithmetic sumset labelling* of a graph G is a modular sumset labelling f of G, with respect to which the set-label of all vertices and edges of G are arithmetic progressions. A graph that admits an arithmetic modular sumset labelling is called an *arithmetic modular sumset graph*.

Analogous to the condition for an arithmetic sumset graphs (see [44]), a necessary and sufficient condition for a graph to admit an arithmetic modular sumset labelling is discussed in the following theorem.

Theorem 3.2. *A graph G admits an arithmetic modular sumset labelling f if and only if for any two adjacent vertices in G, the deterministic ratio of every edge of G is a positive integer, which is less than or equal to the set-labelling number of its end vertex having smaller common difference.*

Proof. Here, we need to consider the following two cases:

Case 1: First note that if the set-label of two adjacent vertices are arithmetic progressions with the same common difference, say d, then the set-label of the corresponding edge is also an arithmetic progression with the same common difference d. Then, it is clear that a vertex arithmetic modular sumset graph is an arithmetic modular sumset graph if the common differences between any two adjacent vertices of G are the same.

Case 2: Assume that u, v be any two adjacent vertices in G with common differences d_u and d_v respectively such that $d_u \leq d_v$. Also, assume that $f(u) = \{a_r = a + rd_u : 0 \leq r < m\}$ and $f(v) = \{b_s = b + sd_v : 0 \leq s < n\}$. Then, $|f(u)| = m$ and $|f(v)| = n$. Now, arrange the terms of $f(u) + f(v) = f^+(uv)$ in rows and columns as follows. For any $b_s \in f(v)$, $0 \leq s < n$, arrange the terms of $A + b_s$ in $(s+1)$th row in such a way that equal terms of different rows come in the same column of this arrangement. Without loss of generality, assume that $d_v = kd_u$ and $k \leq m$. If $k < m$, then for any $a \in f(u)$ and $b \in f(v)$ we have $a + (b + d_v) = a + b + kd_u < a + b + md_i$. That is, a few final elements of each row of the above arrangement occur as the initial elements of the succeeding row (or rows) and the difference between two successive elements in each row is d_u itself. If $k = m$, then the difference between the final element of each row and the first element of the next row is d_u and the difference between two consecutive elements in each row is d_u. Hence, if $k \leq m$, then $f^+(uv)$ is an arithmetic progression with common difference d_u.

In both cases, note that if the given conditions are true, then f is an arithmetic modular sumset labelling of G.

We prove the converse part by contradiction method. For this, assume that f is an arithmetic modular sumset labelling of G. Let us proceed by considering the following two cases.

Case-1: Assume that d_j is not a multiple of d_i (or d_i is not a multiple of d_j). Without loss generality, let $d_i < d_j$. Then, by division algorithm, $d_j = pd_i + q, 0 < q < d_i$. Then, the differences between any two consecutive terms in $f^+(v_iv_j)$ are not equal. Hence, in this case also, f is not an arithmetic modular sumset labelling, contradiction to the hypothesis. Therefore, $d_i | d_j$.

Case 2: Let $d_j = kd_i$ where $k > m$. Then, the difference between two successive elements in each row is d_i, but the difference between the final element of each row and the first element of the next row is td_i, where $t = k - m + 1 \neq 1$. Hence, f is not an arithmetic modular sumset labelling, a contradiction to the hypothesis. Hence, we have $d_j = kd_i; k \leq m$.

Therefore, from the above cases it can be noted that if a vertex arithmetic modular sumset labelling of G is an arithmetic modular sumset labelling of G, then the deterministic ratio of every edge of G is a positive integer, which is greater than or equal to the set-labelling number of its end vertex having smaller common difference. This completes the proof. □

In the following theorem, we establish a relation between the common differences of the elements of an arithmetic modular sumset graph G.

Theorem 3.3. *If G is an arithmetic modular sumset graph, the greatest common divisor of the common differences of vertices of G and the greatest common divisor of the common differences of the edges of G are equal to the smallest among the common differences of the vertices of G.*

Proof. Let f be an arithmetic modular sumset labelling of G. Then, by Theorem 3.2, for any two adjacent vertices v_i and v_j of G with common differences d_i and d_j respectively, either $d_i = d_j$, or if $d_j > d_i, d_j = kd_j$, where k is a positive integer such that $1 < k \leq |f(v_i)|$.

If the common differences of the elements of G are the same, the result is obvious. Hence, assume that for any two adjacent vertices v_i and v_j of G, $d_j = kd_j, k \leq |f(v_i)|$, where d_i is the smallest among the common differences of the vertices of G. If v_r is another vertex that is adjacent to v_j, then it has the common difference d_r which is equal to either d_i or d_j or ld_j. In all the three cases, d_r is a multiple of d_i. Hence, the greatest common divisor of d_i, d_j, d_r is d_i. Proceeding like this, we have the greatest common divisor of the common differences of the vertices of G is d_i.

Also, by Theorem 3.2, the edge u_iv_j has the common difference d_i. The edge v_jv_k has the common difference d_i, if $d_k = d_i$, or d_j in the other two cases. Proceeding like this, we observe that the greatest common divisor of the common differences of the edges of G is also d_i. This completes the proof. □

The study on the set-labelling number of edges of an arithmetic modular sumset graphs arouses much interest. Analogous to the result on set-labelling number of the edges of an arithmetic sumset graph (see [43]), The set-labelling number of an edge of an arithmetic modular sumset graph G, in terms of the set-labelling numbers of its end vertices, is determined in the following theorem.

Theorem 3.4. *Let G be a graph which admits an arithmetic modular sumset labelling, say f and let d_i and d_j be the common differences of two adjacent vertices v_i and v_j in G. If $|f(v_i)| \geq |f(v_j)|$, then for some positive integer $1 \leq k \leq |f(v_i)|$, the edge v_iv_j has the set-labelling number $|f(v_i)| + k(|f(v_j)| - 1)$.*

Proof. Let f be an arithmetic modular sumset labelling defined on G. For any two vertices v_i and v_j of G, let $f(v_i) = \{a_i, a_i + d_i, a_i + 2d_i, a_i + 3d_i, \dots, a_i + (m-1)d_i\}$ and let $f(v_j) = \{a_j, a_j + d_j, a_j + 2d_j, a_j + 3d_j, \dots, a_j + (n-1)d_j\}$. Here $|f(v_i)| = m$ and $|f(v_j)| = n$.

Let d_i and d_j be the common differences of the vertices v_i and v_j respectively, such that $d_i < d_j$. Since f is an arithmetic modular sumset labelling on G, by Theorem 3.2, there exists a positive integer k such that $d_j = k.d_i$, where $1 \leq k \leq |f(v_i)|$. Then, $f(v_j) = \{a_j, a_j + kd_i, a_j + 2kd_i, a_j + 3kd_i, \dots, a_j + (n-1)kd_i\}$. Therefore, $f^+(v_iv_j) = \{a_i + a_j, a_i + a_j + d_i, a_i + a_j + 2d_i, \dots, a_i + a_j + [(m-1) + k(n-1)]d_i\}$. That is, the set-labelling number of the edge v_iv_j is $m + k(n-1)$. □

4. Strongly modular sumset graphs

The next type of a modular sumset labelling we are going to discuss is the one with the upper bound in Inequality (1) is sharp (that is, $|A + B| = |A||B|$). Thus, we have the following definition.

Definition 4.1. [47] A modular sumset labelling $f : V(G) \to \mathcal{P}_0(\mathbb{Z}_n)$ defined on a given graph G is said to be a *strongly modular sumset labelling* if for the associated function $f^+ : E(G) \to \mathcal{P}_0(\mathbb{Z}_n)$, $|f^+(uv)| = |f(u)||f(v)| \forall uv \in E(G)$. A graph which admits a strongly modular sumset labelling is called a *strongly modular sumset graph*.

Invoking the notion difference set of a set, a necessary and sufficient condition of a modular sumset labelling of a graph G to be a strongly modular sumset labelling is given below:

Theorem 4.1. *A modular sumset labelling $f : V(G) \to \mathcal{P}_0(\mathbb{Z}_n)$ of a given graph G is a strongly modular sumset labelling of G if and only if $D_{f(u)} \cap D_{f(v)} = \emptyset, \forall uv \in E(G)$, where $|f(u)||f(v)| \leq n$.*

Proof. Let $f : V(G) \to \mathcal{P}_0(\mathbb{Z}_n)$ be a modular sumset labelling on a given graph G. For any vertex $u \in V(G)$, define $D_f(u) = \{a_i - a_j : a_i, a_j \in f(u)\}$.

Let uv be an arbitrary edge in $E(G)$. Assume that f is a strong modular sumset labelling of G. Then, by definition $|f^+(uv)| = |f(u)||f(v)|$. Therefore, for any elements $a_i, a_j \in f(u)$ and $b_r, b_s \in f(v)$, we have $a_i + b_r \neq a_j + b_s$ in $f^+(uv) \forall uv \in E(G)$. That is, $|a_i - a_j| \neq |b_s - b_r|$ for any $a_i, a_j \in f(u)$ and $b_r, b_s \in f(v)$. That is, $D_{f(u)} \cap D_{f(v)} = \emptyset$. Therefore, the difference sets of the set-label of any two adjacent vertices are disjoint.

Conversely, assume that the difference $D_{f(u)} \cap D_{f(v)} = \emptyset$ for any edge uv in G. That is, $|a_i - a_j| \neq |b_s - b_r|$ for any $a_i, a_j \in f(u)$ and $b_r, b_s \in f(v)$. Then, $a_i - a_j \neq b_s - b_r$. That is, $a_i + b_r \neq a_j + b_s$. Therefore, all elements in $f(u) + f(v)$ are distinct. That is, $|f^+(uv)| = |f(u)||f(v)|$ for any edge $uv \in E(G)$. Hence, f is a strongly modular sumset labelling of G.

Also, note that the maximum possible cardinality in the set-label of any element of G is n, the product $|f(u)||f(v)|$ cannot exceed the number n. This completes the proof. □

A necessary and sufficient condition for a modular sumset labelling of a graph G to be a strongly k-uniform modular sumset labelling is given below:

Theorem 4.2. [47] *For a positive integer $k \leq n$, a modular sumset labelling $f : V(G) \to \mathcal{P}_0(\mathbb{Z}_n)$ of a given connected graph G is a strongly k-uniform modular sumset labelling of G if and only if either k is a perfect square or G is bipartite.*

Proof. If k is a perfect square, say $k = l^2$, then we can label all the vertices of a graph by distinct l-element sets in such a way that the difference sets of the set-label of every pair of adjacent vertices are disjoint. Hence, assume that k is not a perfect square.

Let G be a bipartite graph with bipartition (X, Y). Let r, s be two divisors of k. Label all vertices of X by distinct r-element sets all of whose difference sets are the same, say D_X. Similarly, label all vertices of Y by distinct s-element sets all of whose difference sets the same, say D_Y, such that $D_X \cap D_Y = \emptyset$. Then, all the edges of G have the set-labelling number $k = rs$. Therefore, G is a strongly k-uniform modular sumset graph.

Conversely, assume that G admits a strongly k-uniform modular sumset labelling, say f. Then, $f^+(uv) = k \forall uv \in E(G)$. Since, f is a strong modular sumset labelling, the set-labelling number of every vertex of G is a divisor of the set-labelling numbers of the edges incident on that vertex. Let v be a vertex of G with the set-labelling number r, where r is a divisor of k, but $r^2 \neq k$. Since f is k-uniform, all the vertices in $N(v)$, must have the set-labelling number s, where $rs = k$. Again, all vertices, which are adjacent to the vertices of $N(v)$, must have the set-labelling number r. Since G is a connected graph, all vertices of G have the set-labelling number r or s. Let X be the set of all vertices of G having the set-labelling number r and Y be the set of all vertices of G having the set-labelling number s. Since $r^2 \neq k$, no two elements in X (and no elements in Y also) can be adjacent to each other. Therefore, G is bipartite. □

The following result is an immediate consequence of the above theorem.

Theorem 4.3. [47] *For a positive non-square integer $k \leq n$, a modular sumset labelling $f : V(G) \to \mathcal{P}_0(\mathbb{Z}_n)$ of an arbitrary graph G is a strongly k-uniform modular sumset labelling of G if and only if either G is bipartite or a disjoint union of bipartite components.*

For a positive integer $k \leq n$, the maximum number of components in a strongly k-uniform modular sumset graph is as follows.

Proposition 4.4. [47] *Let f be a strongly k-uniform modular sumset labelling of a graph G with respect to the ground set $\mathbb{Z}_n$. Then, the maximum number of components in G is the number of distinct pairs of divisors r and s of k such that $rs = k$.*

The following theorem discusses the condition for an arithmetic modular sumset labelling of a graph G to be a strongly modular sumset labelling of the graph.

Theorem 4.5. *Let G be a graph which admits an arithmetic modular sumset labelling, say f. Then, f is a strongly modular sumset labelling of G if and only if the deterministic ratio of every edge of G is equal to the set-labelling number of its end vertex having smaller common difference.*

Proof. Let f be an arithmetic modular sumset labelling of G. Let v_i and v_j are two adjacent vertices in G and d_i and d_j be their common differences under f. Without loss of generality, let $d_i < d_j$. Then, by Theorem 3.4, the set-labelling number of the edge v_iv_j is $|f(v_i)| + k(|f(v_j)| - 1)$.

Assume that f is a strongly modular sumset labelling. Therefore, $f^+(v_iv_j) = mn$. Then,

$$
\begin{aligned}
|f(v_i)| + k(|f(v_j)| - 1) &= |f(v_i)||f(v_j)| \\
\Rightarrow k(|f(v_j)| - 1) &= |f(v_i)|(|f(v_j)| - 1) \\
\Rightarrow k &= |f(v_i)|.
\end{aligned}
$$

Conversely, assume that the common differences d_i and d_j of two adjacent vertices v_i and v_j respectively in G, where $d_i < d_j$ such that $d_j = |f(v_i)|.d_i$. Assume that $f(v_i) = \{a_r = a + rd_i : 0 \leq r < |f(v_i)|\}$ and $f(v_j) = \{b_s = b + skd_i : 0 \leq s < |f(v_j)|\}$, where $k \leq |f(v_i)|$. Now, arrange the terms of $f^+(v_iv_j) = f(v_i) + f(v_j)$ in rows and columns as follows. For $b_s \in f(v_j), 0 \leq s < |f(v_j)|$, arrange the terms of $f(v_i) + b_s$ in $(s+1)$-th row in such a way that equal terms of different rows come in the same column of this arrangement. Then, the common difference between consecutive elements in each row is d_i. Since $k = |f(v_i)|$, the difference between the final element of any row (other than the last row) and first element of its succeeding row is also d_i. That is, no column in this arrangement contains more than one element. Hence, all elements in this arrangement are distinct. Therefore, total number of elements in $f(v_i) + f(v_j)$ is $|f(v_i)||f(v_j)|$. Hence, f is a strongly modular sumset labelling. □

5. Supreme modular sumset labelling of *G*

In both types of modular sumset labelling discussed above, it is observed that the cardinality of the edge set-label cannot exceed the value n. This fact creates much interest in investigating the case where all the edge set-label have the cardinality n.

Definition 5.1. [47] A modular sumset labelling $f : V(G) \to \mathcal{P}_0(\mathbb{Z}_n)$ of a given graph G is said to be a *supreme modular sumset labelling* or *maximal modular sumset labelling* of G if and only if $f^+(E(G)) = \{\mathbb{Z}_n\}$.

Put in a different way, a modular sumset labelling $f : V(G) \to \mathcal{P}(\mathbb{Z}_n)$ of a given graph G is a supreme modular sumset labelling of G if the set-label of every edge of G is the ground set $\mathbb{Z}_n$ itself.

A necessary and sufficient condition for a modular sumset labelling of a graph G to be its supreme modular sumset labelling is discussed in the theorem given below:

Theorem 5.1. [47] *The modular sumset labelling $f : V(G) \to \mathcal{P}(\mathbb{Z}_n)$ of a given graph G is a supreme modular sumset labelling of G if and only if for every pair of adjacent vertices u and v of G some or all of the following conditions hold.*

i. $|f(u)| + |f(v)| \geq n$ if $D_{f(u)} \cap D_{f(v)} \neq \emptyset$. The strict inequality hold when $D_{f(u)}$ and $D_{f(v)}$ are arithmetic progressions containing the same elements.

ii. $|f(u)||f(v)| \geq n$ if $D_{f(u)} \cap D_{f(v)} = \emptyset$.

Proof. For two adjacent vertices u and v in G, let $D_{f(u)} = D_{f(v)} = \{d\}$ are arithmetic progressions containing the same elements. Then, the elements in $f(u)$ and $f(v)$ are also in arithmetic progression, with the same common difference d. Then, by Theorem 3.4, $n = |f(u) + f(v)| = |f(u)| + |f(v)| - 1$. Therefore, the set-labelling number of the edge uv is n if and only if $|f(u)| + |f(v)| > n$.

Now, let $D_{f(u)} \cap D_{f(v)} \neq \emptyset$ such that $D_{f(u)} \neq D_{f(v)}$. Then, clearly $|f(u) + f(v)| \geq |f(u)| + |f(v)|$. Therefore, we have $|f^+(uv)| = n$ if and only if $|f(u)| + |f(v)| \geq n$. □

Next assume that $D_{f(u)} \cap D_{f(v)} = \emptyset$. Then, $|f(u) + f(v)| = |f(u)||f(v)|$. Therefore, we have $|f^+(uv)| = n$ if and only if $|f(u)||f(v)| \geq n$.

A necessary and sufficient condition for a strong modular sumset labelling of a graph G to be a maximal modular sumset labelling of G.

Theorem 5.2. [47] *Let f be a strong sumset-labelling of a given graph G. Then, f is a maximal sumset-labelling of G if and only if n is a perfect square or G is bipartite or a disjoint union of bipartite components.*

Proof. The proof is an immediate consequence of Theorem 4.2, when $k = n$. □

6. Weakly modular sumset graphs

Another interesting question we address in the beginning of this section is whether the lower bound and the upper bound of the sumset can be equal. Suppose that A and B be two non-empty subsets of $\mathbb{Z}_n$ such that the bounds of their sumset are equal. Then, we have

$$|A| + |B| - 1 = |A||B|$$
$$|A| + |B| - 1 - |A||B| = 0$$
$$|A|(1-|B|) + |B| - 1 = 0$$
$$(|A|-1)(|B|-1) = 0$$

which is possible only when $|A| = 1$ or $|B| = 1$ (or both). Also, note that in this case the cardinality of the sumset is equal to equal to that of one of the summands. This interesting phenomenon leads us to a new type of a modular sumset labelling called weakly modular sumset labelling. This type of labelling is investigated in the following section.

6.1 Weakly modular sumset labelling of graphs

Definition 6.1. A modular sumset labelling f of a graph G is said to be a *weakly modular sumset labelling* of G if the cardinality of the set-label of every edge of G is equal to the cardinality of the set-label of at least one of its end vertices. A graph which admits a weakly modular sumset labelling is called a *weakly modular sumset graph*.

From the above definition, it can be observed that for any edge uv in weakly modular sumset graph G, $|f^+(uv)| = |f(u)|$ or $|f^+(uv)| = |f(v)|$. Putting it in a different way, the set-labelling number of at least one end vertex of every edge of a weakly modular sumset graph is a singleton. An element (a vertex or an edge) of modular sumset graph G with set-labelling number 1 is called a *sparing element* or a *monocardinal elements* of G. Hence, analogous to the condition for a sumset graph to be a weak sumset graph (see [39]), we have.

Theorem 6.1. *A graph G admits a weak modular sumset labelling if and only if G is bipartite or contains sparing edges.*

Proof. Note the fact that at least one end vertex of every edge of G is a sparing vertex. Also, we note that no two vertices with non-singleton set-label in weakly modular sumset graph can be adjacent to each other. Thus, if every of edge of G has exactly one end vertex with singleton set-label, then we can partition the vertex set of G into two subsets X with all sparing vertices and Y with all non-sparing vertices. Here, no two vertices in the same partition are adjacent and hence G is a bipartite graph. If G is not a bipartite graph, then obviously G should have at least one sparing edge, completing the proof.

Invoking Theorem 6.1, the following two results are immediate.

Corollary 6.2. *Every graph G admits a weakly modular sumset labelling.*

Corollary 6.3. *A graph G admits a weakly uniform modular sumset labelling if and only if G is bipartite.*

The above results are similar to the corresponding result of integer additive set-labelled graphs (see [39]) and hence the notion of *sparing number* of graphs defined and studied in [48–57] can be extended to our current discussion also. The notion of the sparing number of graphs is defined as follows:

Definition 6.2. Let G be a weakly modular sumset labelled graph. Then, the *sparing number* of G is the number of sparing edges in G.

A set of vertices X of a graph G is said to have maximal incidence if the maximum number of edges of incidence at the elements of X. Then, analogous to the corresponding result of integer additive set-valued graphs (see [40]), we have.

Theorem 6.4. *Let G be a weakly modular sumset labelled graph and I be I be the largest independent set of G with maximum incidence. Then, the sparing number of G is* $|E(G)| - \sum_{v_i \in I} d(v_i)$.

Proof. Recall that the degree of a vertex v, denoted by $d(v)$, is equal to the number of edges incident on a vertex. Note that any vertex $v_i \in I$ can have a non-singleton set-label which gives non-singleton set labels to $d(v_i)$ edges incident on it. Since I is an independent set, the edges incident at the vertices in I assumes non-singleton set-label. Therefore, the number of edges having non-singleton set-label incident at the vertices in I is $\sum_{v_i \in I} d(v_i)$. Since I is a maximal independent set of that kind, the above expression counts the maximal non-sparing edges in G. Hence, the number of sparing edges in G is $|E(G)| - \sum_{v_i \in I} d(v_i)$. □

6.2 Weakly modular sumset number of graphs

As a special case of the modular sumset number, the notion of weakly modular sumset number is introduced in [47] as follows:

Definition 6.3. The *weakly modular sumset number* of a graph G, denoted by σ_w is defined to be the minimum value of n such that a modular sumset labelling $f : V(G) \to \mathcal{P}_0(\mathbb{Z}_n)$ is a weakly modular sumset labelling of G.

The following theorem discussed the weak sumset number of an arbitrary graph G in terms of its covering and independence numbers.

Theorem 6.5. [47] *Let G be a modular sumset graph and α and β be the covering number and independence number of G respectively. Then, the weak modular sumset number of G is* $\max\{\alpha, r\}$, *where r is the smallest positive integer such that* $2^r - r - 1 \geq \beta$.

Proof. Recall that $\alpha(G) + \beta(G) = |V(G)|$ (see [4]). Since G is a modular sumset graph, no two adjacent vertices can have non-singleton set-label simultaneously. Therefore, the maximum number of vertices that have non-singleton set-label is β. Let V' be the set of these independent vertices in G. Therefore, the minimum number of sparing vertices is $|V(G)| - \beta = \alpha$. Since all these vertices in $V - V'$ must have distinct singleton set-label, the ground set must have at least α elements.

Also, the number non-empty, non-singleton subsets of the ground set must be greater than or equal to α. Otherwise, all the vertices in V' cannot be labelled by non-singleton subsets of this ground set. We know that the number of non-empty, non-singleton subsets of a set A is $2^{|A|} - |A| - 1$, where $A \subseteq \mathbb{Z}_n$ is the ground set used for labelling.

Therefore, the weak modular sumset number G is α if $2^\alpha - \alpha - 1 \geq \beta$. Otherwise, the ground set must have at least r elements such that $2^r - r - 1 \geq \beta$. Therefore, in this case, the weak modular sumset number of G is r, where r is the smallest positive integer such that $2^r - r - 1 \geq \beta$. Hence, $\sigma^*(G) = \max\{\alpha, r\}$. This completes the proof. □

The weakly modular sumset number some fundamental graph classes are given in **Table 1**.

The following theorem discusses the minimum cardinality of the ground set when the given graph G admits a weakly uniform modular sumset labelling.

Theorem 6.6. [47] *Let G be a weakly k-uniform modular sumset graph with covering number α and independence number β, where $k < \alpha$. Then, the minimum cardinality of the ground set $\mathbb{Z}_n$ is* $\max\{\alpha, r\}$, *where r is the smallest positive integer such that* $\binom{r}{k \geq \beta}$.

Graph class	$\sigma^*(G)$
Path, P_p	2 if $p \leq 2$; $\lfloor \frac{p}{2} \rfloor$ if $p > 2$
Cycle, C_p	$p-1$ if $p = 3, 4$; $\lfloor \frac{p}{2} \rfloor$ if $p > 4$
Wheel graph, $W_{1,p}$	$1 + \lfloor \frac{p}{2} \rfloor$
Helm graph, H_p	p
Ladder graph, L_p	p
Complete graph, K_p	$p-1$

Table 1.
Weakly modular sumset number of some graph classes.

Proof. Let a weakly k-uniform modular sumset labelling be defined on a graph G over the ground set $A \subset \mathbb{Z}_n$. Then, by Corollary 6.3, G is bipartite. Let X, Y be the bipartition of the vertex set $V(G)$. Without loss of generality, let $|X| \leq |Y|$. Then, $\alpha = |X|$ and $\beta = |Y|$. Then, distinct elements of X must have distinct singleton set-label. Therefore, $n \geq \alpha$.

On the other hand, since f is k-uniform, all the elements in Y must have distinct k-element set-label. The number of k-element subsets of a set A (obviously, with more than k elements) is $\binom{|A|}{k}$. The ground set A has α elements only if $\binom{\alpha}{k \geq \beta}$. Otherwise, the ground set A must contain at least r elements, where $r > \alpha$ is the smallest positive integer such that $\binom{r}{k \geq \beta}$. Therefore, $n = \max\{\alpha, r\}$. □

In view of the above theorem, the following result is immediate.

Corollary 6.7. *Let G be a weakly k-uniform modular sumset graph, where $k \geq \alpha$, where α is the covering number of G. Then, the minimum cardinality of the ground set $\mathbb{Z}_n$ is the smallest positive integer n such that $\binom{n}{k \geq \beta}$, where β is the independence number of G.*

The following result explains a necessary and sufficient condition for a weak modular sumset labelling of a given graph G to be a maximal modular sumset labelling of G.

Proposition 6.8. [47] *A weakly modular sumset labelling of a graph G is a supreme modular sumset labelling of G if and only if G is a star graph.*

Proof. Let f be a weak modular sumset labelling of given graph G. First, assume that f is a maximal modular sumset labelling of G. Then, the set-labelling number of one end vertex of every edge of G is 1 and the set-labelling number of the other end vertex is n. Therefore, $\mathbb{Z}_n$ be the set-label of one end vertex of every edge of G, which is possible only if G is a star graph with the central vertex has the set-label $\mathbb{Z}_n$ and the pendant vertices of G have distinct singleton set-label.

Conversely, assume that G is a star graph. Label the central vertex of G by the ground set $\mathbb{Z}_n$ and label other vertices of G by distinct singleton subsets of $\mathbb{Z}_n$. Then, all the edges of G has the set-labelling number n. That is, this labelling is a supreme modular sumset labelling of G. □

7. Conclusion

In this chapter, we have discussed certain types of modular sumset graphs and their structural properties and characterisations. These studies are based on the cardinality of the set-label of the elements of the graphs concerned and the patterns of the elements in these set-label. It is to be noted that several other possibilities can be investigated in this regard. For example, analogous to the topological set-valuations of graphs, the case when the collection of set-label of vertices and/or edges of a graph G forms a topology of the ground set $\mathbb{Z}_n$ can be studied in detail. Another possibility for future investigation is to extend the graceful and sequential concepts of set-labelling of graphs to modular sumset labelling also. All these points highlight the wide scope for further studies in this area.

Acknowledgements

The author would like to dedicate the chapter to his doctoral thesis supervisor Prof. Dr. Germina K Augusthy for her constant support and inspiration for working in the area of sumset-labelling. The author would also like to acknowledge the critical and creative suggestions of the editor and referees which improved the content and the presentation style of the chapter.

Additional information

Mathematics Subject Classification 2010: 05C75

Author details

Sudev Naduvath
Department of Mathematics, CHRIST (Deemed to be University), Bangalore, Karnataka, India

*Address all correspondence to: sudev.nk@christuniversity.in

References

[1] Bondy JA, Murty USR. Graph Theory with Applications. Vol. 290. London: Macmillan; 1976

[2] Bondy JA, Murty USR. Graph Theory. New York: Springer; 2008

[3] Gross JL, Yellen J, Zhang P. Handbook of Graph Theory. Boca Raton: Chapman and Hall/CRC; 2013

[4] Harary F. Graph Theory. New Delhi: Narosa Publ. House; 2001

[5] West DB. Introduction to Graph Theory. Vol. 2. Upper Saddle River: Prentice Hall; 2001

[6] Bača M, MacDougall J, Miller M, Wallis W. Survey of certain valuations of graphs. Discussiones Mathematicae Graph Theory. 2000;**20**(2):219-229

[7] Brandstadt A, Spinrad JP. Graph Classes: A Survey. Vol. 3. Philadelphia: SIAM; 1999

[8] Gallian JA. A dynamic survey of graph labeling. Electronic Journal of Combinatorics 1 (Dynamic Surveys). 2018:DS6

[9] Hammack R, Imrich W, Klavžar S. Handbook of Product Graphs. Bocca Raton: CRC Press; 2011

[10] Imrich W, Klavzar S. Product Graphs: Structure and Recognition. New York: Wiley; 2000

[11] Rosa A. On certain valuations of the vertices of a graph. In: Theory of Graphs International Symposium; July; Rome. 1966. pp. 349-355

[12] Golomb SW. How to number a graph. In: Graph Theory and Computing. Cambridge, UK: Academic Press; 1972. pp. 23-37

[13] Apostol TM. Introduction to Analytic Number Theory. New Delhi: Narosa Publ. House; 2013

[14] Burton DM. Elementary Number Theory. Noida, India: Tata McGraw-Hill Education; 2006

[15] Cohn H. Advanced Number Theory. N Chelmford: Courier Corporation; 2012

[16] Nathanson MB. Elementary Methods in Number Theory. Vol. 195. New York: Springer Science & Business Media; 2008

[17] Acharya BD. Set valuations of a graph and their applications. In: MRI Lecture Notes in Applied Mathematics, 2. Allahabad: Mehta Research Instt; 1986

[18] Acharya BD. Set-indexers of a graph and set-graceful graphs. Bulletin of the Allahabad Mathematical Society. 2001; **16**:1-23

[19] Acharya BD, Germina KA. Set-valuations of graphs and their applications: A survey. Annals of Pure and Applied Mathematics. 2013;**4**(1):8-42

[20] Acharya BD, Hegde SM. On certain vertex valuations of a graph I. Indian Journal of Pure and Applied Mathematics. 1991;**22**(7):553-560

[21] Acharya BD, Rao SB, Arumugam S. Embeddings and NP-complete problems for graceful graphs. In: Labeling of Discrete Structures and Applications. New Delhi: Narosa Pub. House; 2008. pp. 57-62

[22] Augusthy GK. Set-valuations of graphs and their applications. In: Handbook of Research on Advanced Applications of Graph Theory in Modern Society. Hershey, Pennsylvania: IGI Global; 2020. pp. 171-207

[23] Germina KA. Set-valuations of graphs and their applications. Final Technical Report. Vol. 4. DST Grant-in-Aid Project No. SR/S4/277/05

[24] Naduvath S. On disjunctive and conjunctive set-labelings of graphs. Southeast Asian Bulletin of Mathematics. 2019;**43**(4):593-600

[25] Nathanson MB. Additive Number Theory: Inverse Problems and the Geometry of Sumsets. Vol. 165. New York: Springer Science & Business Media; 1996

[26] Nathanson MB. Sums of finite sets of integers. American Mathematical Monthly. 1972;**79**(9):1010-1012

[27] Nathanson MB. Additive Number Theory: The Classical Bases. Vol. 164. New York: Springer Science & Business Media; 2013

[28] Nathanson MB. Problems in Additive Number Theory. I. In Additive Combinatorics. Vol. 43. Providence, RI: American Mathematical Society; 2007. pp. 263-270

[29] Ruzsa IZ. Generalized arithmetical progressions and sumsets. Acta Mathematica Hungarica. 1994;**65**(4): 379-388

[30] Ruzsa IZ. Sumsets and structure. In: Combinatorial Number Theory and Additive Group Theory. 2009. pp. 87-210

[31] Sudev NK, Germina KA, Chithra KP. Weak integer additive set-labeled graphs: A creative review. Asian-European Journal of Mathematics. 2015;**8**(3):1550052

[32] Sudev NK, Germina KA, Chithra KP. Strong integer additive set-valued graphs: A creative review. International Journal of Computer Applications. 2015;**97**(5):8887

[33] Sudev NK, Germina KA, Chithra KP. Arithmetic integer additive set-valued graphs: A creative review. The Journal of Mathematics and Computer Science. 2020;**10**(4): 1020-1049

[34] Tao T. Sumset and inverse sumset theory for Shannon entropy. Combinatorics, Probability and Computing. 2010;**19**(4):603-639

[35] Sudev NK, Germina KA. On integer additive set-indexers of graphs. International Journal of Mathematical Sciences and Engineering Applications. 2015;7(01):1450065

[36] Germina KA, Sudev NK. On weakly uniform integer additive set-indexers of graphs. International Mathematical Forum. 2013;**8**(37):1837-1845

[37] Germina KA, Anandavally TMK. Integer additive set-indexers of a graph: Sum square graphs. Journal of Combinatorics, Information & System Sciences. 2012;**37**(2-4):345-358

[38] Sudev NK. Set valuations of discrete structures and their applications [PhD thesis]. Kannur University; 2015

[39] Naduvath S, Germina K. A characterisation of weak integer additive set-indexers of graphs. Journal of Fuzzy Set Valued Analysis. 2014, 2014;**1** jfsva-00189

[40] Sudev NK, Germina KA, Chithra KP. Weak set-labeling number of certain integer additive set-labeled graphs. International Journal of Computers and Applications. 2015; **114**(2):1-6

[41] Sudev NK, Germina KA. On weak integer additive set-indexers of certain graph classes. Journal of Discrete Mathematical Sciences & Cryptography. 2015;**18**(1–2):117-128

[42] Sudev NK, Germina KA. Some new results on strong integer additive set-indexers of graphs. Discrete Mathematics, Algorithms and Applications. 2015;**7**(01):1450065

[43] Sudev NK, Germina KA. On certain arithmetic integer additive set-indexers of graphs. Discrete Mathematics, Algorithms and Applications. 2015; 7(03):1550025

[44] Sudev NK, Germina KA. A study on arithmetic integer additive set-indexers of graphs. Journal of Informatics and Mathematical Sciences. 2018;**10**(1–2): 321-332

[45] Naduvath S, Augusthy GK, Kok J. Sumset valuations of graphs and their applications. In: Handbook of Research on Advanced Applications of Graph Theory in Modern Society. Hershey, Pennsylvania: IGI Global; 2020. pp. 208-250

[46] Naduvath S, Germina KA. An Introduction to Sumset Valued Graphs. Mauritius: Lambert Academic Publ; 2018

[47] Naduvath S. A study on the modular sumset labeling of graphs. Discrete Mathematics, Algorithms and Applications. 2017;**9**(03):1750039

[48] Chithra KP, Sudev NK, Germina KA. Sparing number of Cartesian products of certain graphs. Communications in Mathematics and Applications. 2014;**5**(1):23-30

[49] Chithra KP, Sudev NK, Germina KA. A study on the sparing number of corona of certain graphs. Research & Reviews: Discrete Mathematical Structures. 2014;**1**(2):5-15

[50] Naduvath S, Kaithavalappil C, Augustine G. A note on the sparing number of generalised petersen graphs. Journal of Combinatorics, Information & System Sciences. 2017;**42**(1–2):23-31

[51] Sudev NK, Germina KA. A note on the sparing number of graphs. Advances and Applications in Discrete Mathematics. 2014;**14**(1):51-65

[52] Sudev NK, Germina KA. On the sparing number of certain graph structures. Annals of Pure and Applied Mathematics. 2014;**6**(2):140-149

[53] Sudev NK, Germina KA. Further studies on the sparing number of graphs. TechS Vidya e-Journal of Research. 2014;**2**(2):25-36

[54] Sudev NK, Germina KA. A note on the sparing number on the sieve graphs of certain graphs. Applied Mathematics E-Notes. 2015;**15**(1):29-37

[55] Sudev NK, Germina KA. Some new results on weak integer additive set-labeled graphs. International Journal of Computers and Applications. 2015; **128**(1):1-5

[56] Sudev NK, Chithra KP, Germina KA. Sparing number of the certain graph powers. Acta Universitatis Sapientiae Mathematica. 2019;**11**(1): 186-202

[57] Sudev NK, Chithra KP, Germina KA. Sparing number of the powers of certain Mycielski graphs. Algebra and Discrete Mathematics. 2019;**28**(2):291-307

Chapter 7

Determination of the Properties of (p,q)-Sigmoid Polynomials and the Structure of Their Roots

Jung Yoog Kang

Abstract

Nowadays, many mathematicians have great concern about (p,q)-numbers, which are various applications, and have studied these numbers in many different research areas. We know that (p,q)-numbers are different to q-numbers because of the symmetric property. We find the addition theorem, recurrence formula, and (p,q)-derivative about sigmoid polynomials including (p,q)-numbers. Also, we derive the relevant symmetric relations between (p,q)-sigmoid polynomials and (p,q)-Euler polynomials. Moreover, we observe the structures of appreciative roots and fixed points about (p,q)-sigmoid polynomials. By using the fixed points of (p,q)-sigmoid polynomials and Newton's algorithm, we show self-similarity and conjectures about (p,q)-sigmoid polynomials.

Keywords: (p,q)-sigmoid numbers, (p,q)-sigmoid polynomials, (p,q)-Euler polynomials, roots structure, fixed point

1. Introduction

In 1991, Chakrabarti and Jagannathan [1] introduced the (p,q)-number in order to unify varied forms of q-oscillator algebras in physics literature. Around the same time, Brodimas et al. and Arik et al. independently discovered the (p,q)-number (see [2, 3]). Contemporarily, Wachs and White [4] introduced the (p,q)-number in mathematics literature by certain combinatorial problems without any connection to the quantum group related to mathematics and physics literature.

For any $n \in \mathbb{C}$, the (p,q)-number is defined by

$$[n]_{p,q} = \frac{p^n - q^n}{p - q}, \qquad \left|\frac{q}{p}\right| < 1. \tag{1}$$

Thereby, several physical and mathematical problems lead to the necessity of (p,q)-calculus. Based on the aforementioned papers, many mathematicians and physicists have developed the (p,q)-calculus in many different research areas (see [1–21]).

Definition 1.1. *Let z be any complex numbers with $|z| < 1$. The two forms of (p,q)-exponential functions are defined by*

$$e_{p,q}(z) = \sum_{n=0}^{\infty} p^{\binom{n}{2}} \frac{z^n}{[n]_{p,q}!},$$
$$e_{p^{-1},q^{-1}}(z) = \sum_{n=0}^{\infty} q^{\binom{n}{2}} \frac{z^n}{[n]_{p,q}!}. \tag{2}$$

The useful relation of two forms of (p,q)-exponential functions is taken by

$$e_{p,q}(z)e_{p^{-1},q^{-1}}(-z) = 1. \tag{3}$$

In [9], Corcino created the theorem of (p,q)-extension of binomials coefficients and found various properties which are related to horizontal function, triangular function, and vertical function.

Definition 1.2. *Let $n \geq k$. (p,q)-Gauss Binomial coefficients are defined by*

$$\begin{bmatrix} n \\ k \end{bmatrix}_{p,q} = \frac{[n]_{p,q}!}{[n-k]_{p,q}![k]_{p,q}!}, \tag{4}$$

where $[n]_{p,q}! = [n]_{p,q}[n-1]_{p,q} \cdots [1]_{p,q}$.

In 2013, Sadjang [21] derived some properties of the (p,q)-derivative, (p,q)-integration and investigated two (p,q)-Taylor formulas for polynomials.

Definition 1.3. *We define the (p,q)-derivative operator of any function f, also referred to as the Jackson derivative, as follows:*

$$D_{p,q}f(x) = \frac{f(px) - f(qx)}{(p-q)x}, \qquad x \neq 0, \tag{5}$$

and $D_{p,q}f(0) = f'(0)$.

If $t(x) = \sum_{k=0}^{n} a_k x^k$ then $D_{p,q}t(x) = \sum_{k=0}^{n-1} a_{k+1}[k+1]_{p,q}x^k$, since $D_{p,q}z^n = [n]_{p,q}z^{n-1}$. This equation is equivalent to the (p,q)-difference equation in q with known f, $D_{p,q}g(x) = f(x)$..

Theorem 1.4. *This operator, $D_{p,q}$, has the following basic properties:*

(i) Derivative of a product
$$\begin{aligned} D_{p,q}(f(x)g(x)) &= f(px)D_{p,q}g(x) + g(qx)D_{p,q}f(x) \\ &= g(px)D_{p,q}f(x) + f(qx)D_{p,q}g(x). \end{aligned} \tag{6}$$

(ii) Derivative of a ratio
$$\begin{aligned} D_{p,q}\left(\frac{f(x)}{g(x)}\right) &= \frac{g(qx)D_{p,q}f(x) - f(qx)D_{p,q}g(x)}{g(px)g(qx)} \\ &= \frac{g(px)D_{p,q}f(x) - f(px)D_{p,q}g(x)}{g(px)g(qx)}. \end{aligned} \tag{7}$$

Let f be an arbitrary function. In [7], we note that the definition of (p,q)-integral is

$$\int f(x)d_{p,q}x = (p-q)x \sum_{k=0}^{\infty} \frac{q^k}{p^{k+1}} f\left(\frac{q^k}{p^{k+1}}x\right). \tag{8}$$

In 2016, Araci et al. [6] introduced a new class of Bernoulli, Euler and Genocchi polynomials based on the theory of (p, q)-number and found some properties including difference equations, addition theorem, recurrence relations were derived. We observe some special properties and roots structures of Bernoulli, Euler, and tangent polynomials (see [4, 11, 16–20]). In particular, roots structures and fixed points of tangent polynomials including q-numbers are shown in a different shape by [17].

Definition 1.5. *(p, q)-Euler polynomials are defined by*

$$\sum_{n=0}^{\infty} E_{n,p,q}(x)\frac{t^n}{[n]_{p,q}!} = \frac{2}{e_{p,q}(t)+1}e_{p,q}(tx). \tag{9}$$

Several studies have investigated the sigmoid function for various applications (see [11, 12, 15, 16]). For example, a variant sigmoid function with three parameters has been employed to explain hybrid sigmoidal networks [10] and sigmoid function has been defined using flexible sigmoidal mixed models based on logistic family curves for medical applications [11, 12, 15].

Definition 1.6. *We define the sigmoid polynomials as follows:*

$$\sum_{n=0}^{\infty} S_n(x)\frac{t^n}{n!} = \frac{1}{e^{-t}+1}e^{tx}. \tag{10}$$

One of the most widely used methods of solving equations is Newton's method. This method is also based on a linear approximation of the function, but does so using a tangent to the curve. Starting from an initial estimate that is not too far from a root x, we extrapolate along the tangent to its intersection with the x-axis, and take that as the next approximation. This is continued until either the successive x-values are sufficiently close, or the value of the function is sufficiently near zero.

The calculation scheme follows immediately from the right triangle, which has the angle of inclination of the tangent line to the curve at $x = x_1$ as one of its acute angles:

$$\tan\theta = f'(x_1) = \frac{f(x_1)}{x_1 - x_2}, \qquad x_2 = x_1 - \frac{f(x_1)}{f'(x_1)}.$$

We continue the calculation scheme by computing

$$x_3 = x_2 - \frac{f(x_2)}{f'(x_2)},$$

or, in more general terms,

$$x_{n+1} = x_n - \frac{f(x_n)}{f'(x_n)}, \qquad n = 1, 2, 3, \cdots.$$

Newton's algorithm is widely used because, at least in the near neighborhood of a root, it is more rapidly convergent than any of the methods so far discussed. The method is quadratically convergent, by which we mean that the error of each step approaches a constant K times the square of the error of the previous step. The net result of this is that the number of decimal places of accuracy nearly doubles at each iteration. However, offsetting this is the need for two function evaluations at each step, $f(x_n)$ and $f'(x_n)$. We now use the result to show a criterion for convergence of

Newton's method. Consider the form $x_{n+1} = g(x_n)$. Successive iterations converge if $|g'(x)| < 1$. Since

$$g(x) = x - \frac{f(x)}{f'(x)},$$
$$g'(x) = 1 - \frac{f'(x)f'(x) - f(x)f''(x)}{(f'(x))^2} = \frac{f(x)f''(x)}{(f'(x))^2}.$$

Hence if

$$\left|\frac{f(x)f''(x)}{(f'(x))^2}\right| < 1$$

on an interval about the root r, the method will converge for any initial value x_1 in the interval. The condition is sufficient only, and requires the unusual continuity and existence of $f(x)$ and its derivatives. Note that $f'(x)$ must not zero. In addition, Newton's method is quadratically convergent and we can apply this method to polynomials.

Let $f : D \to D$ be a complex function, with D as a subset of $\mathbb{C}$. We define the iterated maps of the complex function as the following:

$$f_r : z_0 \to \underbrace{f(f(\cdots(f}_{r}(z_0)\cdots))$$

The iterates of f are the functions $f, f \circ f, f \circ f \circ f, \ldots,$ which are denoted $f^1, f^2, f^3, \ldots$. If $z \in \mathbb{C}$, and then the orbit of z_0 under f is the sequence $< z_0, f(z_0), f(f(z_0)), \cdots >$.

Definition 1.7. *The orbit of the point $z_0 \in \mathbb{C}$ under the action of the function f is said to be bounded if there exists $M \in \mathbb{R}$ such that $|f^n(z_0)| < M$ for all $n \in \mathbb{N}$. If the orbit is not bounded, it is said to be unbounded.*

Definition 1.8. *Let $f : D \to \mathbb{C}$ be a transformation on a metric space. A point $z_0 \in D$ such that $f(z_0) = z_0$ is called a fixed point of the transformation.*

We know that the fixed point is divided as follows. Suppose that the complex function f is analytic in a region D of $\mathbb{C}$, and f has a fixed point at $z_0 \in D$. Then z_0 is said to be:

an attracting fixed point if $|f'(z_0)| < 1$;

a repelling fixed point if $|f'(z_0)| > 1$;

a neutral fixed point if $|f'(z_0)| = 1$.

If z_0 is an attracting fixed point of f, then there exists a neighborhood of A such that if $b \in A$ the orbit b converges to z_0. Attractive fixed points of a function have a basin of attraction, which may be disconnected. The component which contains the fixed point is called the immediate basin of attraction. If z_0 is a repelling periodic point of f, then there is a neighborhood of N such that if $b \in N$, there are points in the orbit of b which are not in N. In the case of polynomials of degree greater than 0 and some rational functions, ∞ is also called an attracting fixed point, as, for each such function, f, there exist $R > 0$ such that if $|z| > R$ then $f^n(z) \to \infty$ as $n \to \infty$.

Based on the above, the contents of the paper are as follows. Section 2 checks the properties of (p,q)-sigmoid polynomials. For example, we look for addition theorem, recurrence relation, differential, etc. and find the properties associated with the symmetric property and (p,q)-Euler polynomials. Section 3 identifies the structure and accumulation of roots of (p,q)-sigmoid polynomials based on the contents of Section 2 and checks the contents related to the fixed points. Also, we use

Newton's method to obtain a iterative function of (p, q)-sigmoid polynomials to identify the domain leading to the fixed points.

2. Some properties and identities of (p, q)-sigmoid polynomials

This section introduces about (p, q)-sigmoid numbers and polynomials. From the generating function of these polynomials, we can observe some of the basic properties and identities of this polynomials. In particular, we can show the forms of (p, q)-derivative, symmetric properties, and relations of (p, q)-Euler polynomials for (p, q)-sigmoid polynomials.

Definition 2.1. *We define* (p, q)*-sigmoid polynomials as following:*

$$\sum_{n=0}^{\infty} \mathcal{S}_{n,p,q}(x)\frac{t^n}{[n]_{p,q}!} = \frac{1}{e_{p,q}(-t)+1}e_{p,q}(tx). \tag{11}$$

In the Definition 2.1, if $x = 0$ we can see that

$$\sum_{n=0}^{\infty} \mathcal{S}_{n,p,q}(0)\frac{t^n}{[n]_{p,q}!} = \sum_{n=0}^{\infty} \mathcal{S}_{n,p,q}\frac{t^n}{[n]_{p,q}!} = \frac{1}{e_{p,q}(-t)+1}, \tag{12}$$

so we can be called $\mathcal{S}_{n,p,q}$ is (p, q)-sigmoid numbers. We note that $e_{p,q}(0) = p^{\binom{n}{2}}$ because of the property for (p, q)-exponential function. If $p = 1$ in the Definition 2.1, then one holds

$$\sum_{n=0}^{\infty} \mathcal{S}_{n,1,q}(x)\frac{t^n}{[n]_{1,q}!} = \sum_{n=0}^{\infty} \mathcal{S}_{n,q}(x)\frac{t^n}{[n]_q!} = \frac{1}{e_q(-t)+1}e_q(tx), \tag{13}$$

where $\mathcal{S}_{n,q}(x)$ is q-sigmoid polynomials. Moreover, if $p = 1, q \to 1$ in the generation function of (p, q)-sigmoid polynomials, we have

$$\lim_{q\to 1}\sum_{n=0}^{\infty} \mathcal{S}_{n,1,q}(x)\frac{t^n}{[n]_{1,q}!} = \sum_{n=0}^{\infty} \mathcal{S}_n(x)\frac{t^n}{n!} = \frac{1}{e^{-t}+1}e^{tx}, \tag{14}$$

where $\mathcal{S}_n(x)$ is sigmoid polynomials (see [16]).

Theorem 2.2. *Let be* $|q/p| < 1$*. Then we get*

$$\begin{aligned}
&(i)\quad p^{\binom{n}{2}}x^n = \sum_{k=0}^{n}\begin{bmatrix} n \\ k \end{bmatrix}_{p,q}(-1)^{n-k}p^{\binom{n-k}{2}}\mathcal{S}_{k,p,q}(x) + \mathcal{S}_{n,p,q}(x),\\
&(ii)\quad \sum_{k=0}^{n}\begin{bmatrix} n \\ k \end{bmatrix}_{p,q}(-1)^{n-k}p^{\binom{n-k}{2}}\mathcal{S}_{k,p,q} + \mathcal{S}_{n,p,q} = \begin{cases} 1 & \text{if} \quad n = 0, \\ 0 & \text{if} \quad n \neq 0. \end{cases}
\end{aligned} \tag{15}$$

Proof. (i) Consider that $e_{p,q}(-t) \neq -1$. Then we can see

$$\sum_{n=0}^{\infty} \mathcal{S}_{n,p,q}(x)\frac{t^n}{[n]_{p,q}!}\left(e_{p,q}(-t)+1\right) = e_{p,q}(tx). \tag{16}$$

Using power series of (p,q)-exponential function in the equation above (16) and Cauchy product, we can compare both-sides as following:

$$\sum_{n=0}^{\infty}\left(\sum_{k=0}^{n}\begin{bmatrix} n \\ k \end{bmatrix}_{p,q}(-1)^{n-k}p^{\binom{n-k}{2}}\mathcal{S}_{k,p,q}(x)+\mathcal{S}_{n,p,q}(x)\right)\frac{t^n}{[n]_{p,q}!}$$
$$=\sum_{n=0}^{\infty}p^{\binom{n}{2}}x^n\frac{t^n}{[n]_{p,q}!}, \tag{17}$$

that is shown the required result of Theorem 2.2 (i).

(ii) This equation is a recurrence formulae of (p,q)-sigmoid numbers. We omit the proof of Theorem 2.2 (ii) since we can find the result for (p,q)-sigmoid numbers to calculating the same method (i). □

Based on the results from Definition 2.1 and Theorem 2.2, can be taken a few (p,q)-sigmoid numbers and polynomials can be calculated by using computer. We can observe that some of the (p,q)-sigmoid numbers are $\mathcal{S}_{0,p,q}=1/2$, $\mathcal{S}_{1,p,q}=1/4(-p+q)$, $\mathcal{S}_{2,p,q}=1/8(p+q)(p^2-3pq+q^2)$, $\mathcal{S}_{3,p,q}=-1/16(p-q)(p+q)(p^2-4pq+q^2)(p^2+pq+q^2)$, $\mathcal{S}_{4,p,q}=1/32(p^2+q^2)(p^8-4p^7q-4p^6q^2+7p^5q^3+8p^4q^4+7p^3q^5-4p^2q^6-4pq^7+q^8)$, $\cdots$.

Example 2.3. *Some of the (p,q)-sigmoid polynomials are:*

$$\begin{aligned}
\mathcal{S}_{0,p,q}(x)&=\frac{1}{2}\\
\mathcal{S}_{1,p,q}(x)&=\frac{1}{4}(1+2x)\\
\mathcal{S}_{2,p,q}(x)&=\frac{1}{8}(-p+q+2(p+q)x+4px^2)\\
\mathcal{S}_{3,p,q}(x)&=\frac{1}{16}(q^3(1+2x)+2p^2q(-1+2x^2)+2pq^2(-1+2x^2)+p^3(1-2x+4x^2+8x^3))\\
\mathcal{S}_{4,p,q}(x)&=\frac{1}{32}(-3p^2q^4(1+2x)+q^6(1+2x)+4p^3q^3x(-1+x+2x^2))\\
&\quad+\frac{1}{32}(p^4q^2(-1+2x)(-3+4x^2)+pq^5(-3-2x+4x^2)+p^5q(3-2x+8x^3))\\
&\quad+\frac{1}{32}(p^6(-1+2x(1-2x+4x^2+8x^3)))\\
&\cdots.
\end{aligned} \tag{18}$$

Theorem 2.4. *Let k be a nonnegative integer. Then we obtain*

$$\begin{aligned}
(i)\quad \mathcal{S}_{n,p,q}(x)&=\sum_{k=0}^{n}\begin{bmatrix} n \\ k \end{bmatrix}_{p,q}p^{\binom{n-k}{2}}\mathcal{S}_{k,p,q}x^{n-k},\\
(ii)\quad \mathcal{S}_{n,p,q}(x,y)&=\sum_{k=0}^{n}\begin{bmatrix} n \\ k \end{bmatrix}_{p,q}p^{\binom{n-k}{2}}\mathcal{S}_{k,p,q}y^{n-k}.
\end{aligned} \tag{19}$$

Proof. (i) Using the definition of (p,q)-exponential function, we can transform the Definition 2.1 as the follows:

$$\sum_{n=0}^{\infty} \mathcal{S}_{n,p,q}(x) \frac{t^n}{[n]_{p,q}!} = \sum_{n=0}^{\infty} \mathcal{S}_{n,p,q} \frac{t^n}{[n]_{p,q}!} \sum_{n=0}^{\infty} p^{\binom{n}{2}} x^n \frac{t^n}{[n]_{p,q}!}$$
$$= \sum_{n=0}^{\infty} \left(\sum_{k=0}^{n} \begin{bmatrix} n \\ k \end{bmatrix}_{p,q} p^{\binom{n-k}{2}} \mathcal{S}_{k,p,q} x^{n-k} \right) \frac{t^n}{[n]_{p,q}!}. \tag{20}$$

Therefore, we complete the proof of the Theorem 2.4 (i) at once.

(ii) We also consider (p, q)-sigmoid polynomials in two parameters. Then we derive

$$\sum_{n=0}^{\infty} \mathcal{S}_{n,p,q}(x,y) \frac{t^n}{[n]_{p,q}!} = \frac{1}{e_{p,q}(-t)+1} e_{p,q}(tx) e_{p,q}(ty)$$
$$= \sum_{n=0}^{\infty} \left(\sum_{k=0}^{n} \begin{bmatrix} n \\ k \end{bmatrix}_{p,q} p^{\binom{n-k}{2}} \mathcal{S}_{k,p,q}(x) y^{n-k} \right) \frac{t^n}{[n]_{p,q}!}, \tag{21}$$

where the required result (ii) is completed immediately. □

Theorem 2.5. *Let* $|q/p| < 1$ *and k be a nonnegative integer. Then we have*

$$(i) \quad \mathcal{S}_{n,p,q}(x) = \sum_{k=0}^{n} \begin{bmatrix} n \\ k \end{bmatrix}_{p,q} (-1)^k p^{\binom{n-k}{2} + \binom{k}{2}} q^{\binom{k}{2}} \mathcal{S}_{n,p^{-1},q^{-1}}(-1) x^{n-k},$$
$$(ii) \quad \mathcal{S}_{n,p,q} = (-1)^n p^{\binom{n}{2}} q^{\binom{n}{2}} \mathcal{S}_{n,p^{-1},q^{-1}}(-1). \tag{22}$$

Proof. (i) Multiplying $e_{p^{-1},q^{-1}}(t)$ in generating function of (p, q)-sigmoid polynomials, we can investigate

$$\sum_{n=0}^{\infty} \mathcal{S}_{n,p,q}(x) \frac{t^n}{[n]_{p,q}!}$$
$$= \frac{1}{1+e_{p^{-1},q^{-1}}(t)} e_{p^{-1}q^{-1}}(t) e_{p,q}(tx)$$
$$= \sum_{n=0}^{\infty} \mathcal{S}_{n,p^{-1},q^{-1}}(-1) \frac{(-t)^n}{[n]_{p^{-1},q^{-1}}!} \sum_{n=0}^{\infty} p^{\binom{n}{2}} x^n \frac{t^n}{[n]_{p,q}!}$$
$$= \sum_{n=0}^{\infty} \left(\sum_{k=0}^{n} \begin{bmatrix} n \\ k \end{bmatrix}_{p,q} (-1)^k p^{\binom{n-k}{2} + \binom{k}{2}} q^{\binom{k}{2}} \mathcal{S}_{n,p^{-1},q^{-1}}(-1) x^{n-k} \right) \frac{t^n}{[n]_{p,q}!}. \tag{23}$$

Comparing the both-side in the equation above, (23), we find the required results (i).

(ii) Using the same method (i), we make the equation (ii), so we omit the proof of Theorem 2.5 (ii). □

Corollary 2.6. *From Theorem 2.5, one holds*

$$\sum_{k=0}^{n}\begin{bmatrix} n \\ k \end{bmatrix}_{p,q} p^{\binom{n-k}{2}} \mathcal{S}_{k,p,q} = \sum_{k=0}^{n}\begin{bmatrix} n \\ k \end{bmatrix}_{p,q} (-1)^k p^{\binom{n-k}{2}+\binom{k}{2}} q^{\binom{k}{2}} \mathcal{S}_{n,p^{-1},q^{-1}}(-1). \tag{24}$$

Theorem 2.7. *For $|q/p|<1$, (p,q)-derivative of (p,q)-sigmoid polynomials is as the follows:*

$$\frac{\mathcal{D}_{p,q}}{\mathcal{D}_{p,q}x}\mathcal{S}_{n,p,q}(x) = [n]_{p,q}\mathcal{S}_{n-1,p,q}(px). \tag{25}$$

Proof. Using (p,q)-derivative for $\mathcal{S}_{n,p,q}(x)$, we have

$$\begin{aligned}\sum_{n=0}^{\infty}\frac{\mathcal{D}_{p,q}}{\mathcal{D}_{p,q}x}\mathcal{S}_{n,p,q}(x)\frac{t^n}{[n]_{p,q}!} &= \frac{1}{e_{p,q}(-t)+1}\mathcal{D}_{p,q}e_{p,q}(tx) \\ &= \sum_{n=0}^{\infty}\mathcal{S}_{n,p,q}\frac{t^n}{[n]_{p,q}!}\sum_{n=0}^{\infty}p^{\binom{n}{2}}\mathcal{D}_{p,q}x^n\frac{t^n}{[n]_{p,q}!}.\end{aligned} \tag{26}$$

Here, we can note that $\mathcal{D}_{p,q}x^n = \frac{(px)^n-(qx)^n}{(p-q)x} = [n]_{p,q}x^{n-1}$ (see [7]). From the equation above (26), we can transform the equation to

$$\begin{aligned}&\sum_{n=0}^{\infty}\frac{\mathcal{D}_{p,q}}{\mathcal{D}_{p,q}x}\mathcal{S}_{n,p,q}(x)\frac{t^n}{[n]_{p,q}!} \\ &= \sum_{n=0}^{\infty}\left(\sum_{k=0}^{n}\begin{bmatrix} n \\ k \end{bmatrix}_{p,q}[n-k]_{p,q}p^{\binom{n-k}{2}}\mathcal{S}_{k,p,q}x^{n-1-k}\right)\frac{t^n}{[n]_{p,q}!} \\ &= \sum_{n=0}^{\infty}\left(\sum_{k=0}^{n-1}[n]_{p,q}\begin{bmatrix} n-1 \\ k \end{bmatrix}_{p,q}p^{\binom{n-1-k}{2}}\mathcal{S}_{k,p,q}(px)^{n-1-k}\right)\frac{t^n}{[n]_{p,q}!}.\end{aligned} \tag{27}$$

Using the comparison of coefficients in the both-sides, we can find

$$\frac{\mathcal{D}_{p,q}}{\mathcal{D}_{p,q}x}\mathcal{S}_{n,p,q}(x) = [n]_{p,q}\sum_{k=0}^{n-1}\begin{bmatrix} n-1 \\ k \end{bmatrix}_{p,q}p^{\binom{n-1-k}{2}}\mathcal{S}_{k,p,q}(px)^{n-1-k}. \tag{28}$$

Applying the Theorem 2.4 (i) in the equation above, (28), we complete the proof of Theorem 2.7. □

Theorem 2.8. *Let $|q/p|<1$ and $p \neq q$. Then we investigate*

$$\mathcal{D}_{p,q}\mathcal{S}_{n,p,q}(x) = \frac{\mathcal{S}_{n,p,q}(px) - \mathcal{S}_{n,p,q}(qx)}{(p-q)x}. \tag{29}$$

Proof. Applying (p,q)-derivative in the (p,q)-exponential function, we have

$$\begin{aligned}\mathcal{D}_{p,q}\sum_{n=0}^{\infty}\mathcal{S}_{n,p,q}(x)\frac{t^n}{[n]_{p,q}!} &= \sum_{n=0}^{\infty}\mathcal{D}_{p,q}\mathcal{S}_{n,p,q}(x)\frac{t^n}{[n]_{p,q}!}\\ &= \frac{1}{e_{p,q}(-t)+1}\left(\frac{e_{p,q}(ptx)-e_{p,q}(qtx)}{(p-q)x}\right)\\ &= \frac{1}{(p-q)x}\sum_{n=0}^{\infty}\left(\mathcal{S}_{n,p,q}(px)-\mathcal{S}_{n,p,q}(qx)\right)\frac{t^n}{[n]_{p,q}!},\end{aligned} \tag{30}$$

which is the required result. □

Corollary 2.9. *Comparing the Theorem 2.7 and Theorem 2.8, one holds*

$$[n]_{p,q}(p-q)x\mathcal{S}_{n-1,p,q}(px) + \mathcal{S}_{n,p,q}(qx) = \mathcal{S}_{n,p,q}(px). \tag{31}$$

Corollary 2.10. *Putting $x = 1$ in the Theorem 2.7 and Theorem 2.8, one holds*

$$[n]_{p,q}(p-q)\sum_{k=0}^{n-1}\begin{bmatrix} n-1 \\ k \end{bmatrix}_{p,q} p^{\binom{n-1-k}{2}+n-1-k}\mathcal{S}_{k,p,q} = \mathcal{S}_{n,p,q}(p) - \mathcal{S}_{n,p,q}(q). \tag{32}$$

Theorem 2.11. *Let $|q/p|<1$, $p \neq q$, and $p,\ q \neq 0$. Then we obtain*

$$p\left(pqxD\mathcal{S}_{n,p,q}(x) + \mathcal{D}_{p,q}\mathcal{S}_{n,p,q}(qx)\right) = q\left(\mathcal{D}_{p,q}\mathcal{S}_{n,p,q}(x) + pqx\mathcal{D}^{(2)}_{p,q}\mathcal{S}_{n,p,q}(x)\right). \tag{33}$$

Proof. Using the Theorem 2.8 above, we have

$$\sum_{n=0}^{\infty}\mathcal{D}^{(2)}_{p,q}\mathcal{S}_{n,p,q}(x)\frac{t^n}{[n]_{p,q}!} = \frac{1}{(p-q)\left(e_{p,q}(-t)+1\right)}\left(\frac{e_{p,q}(ptx)-e_{p,q}(qtx)}{x}\right). \tag{34}$$

Here, we can obtain that

$$\begin{aligned}(i)\quad \mathcal{D}_{p,q}\frac{e_{p,q}(ptx)}{x} &= \frac{1}{pqx^2}\left(qx\mathcal{D}_{p,q}e_{p,q}(ptx) - e_{p,q}(pqtx)\mathcal{D}_{p,q}x\right)\\ &= \frac{qe_{p,q}(p^2tx) - pe_{p,q}(pqtx)}{pq(p-q)x^2},\end{aligned} \tag{35}$$

and

$$\begin{aligned}(ii)\quad \mathcal{D}_{p,q}\frac{e_{p,q}(qtx)}{x} &= \frac{1}{pqx^2}\left(qx\mathcal{D}_{p,q}e_{p,q}(qtx) - e_{p,q}(q^2tx)\mathcal{D}_{p,q}x\right)\\ &= \frac{qe_{p,q}(pqtx) - pe_{p,q}(q^2tx)}{pq(p-q)x^2}.\end{aligned} \tag{36}$$

Applying Eqs. (35) and (36) in the Eq. (34), we can catch the following equation:

$$\sum_{n=0}^{\infty} \mathcal{D}_{p,q}^{(2)} \mathcal{S}_{n,p,q}(x) \frac{t^n}{[n]_{p,q}!} = \frac{1}{(p-q)pqx} \left(\sum_{n=0}^{\infty} \left(q\mathcal{D}_{p,q}\mathcal{S}_{n,p,q}(px) - p\mathcal{D}_{p,q}\mathcal{S}_{n,p,q}(qx) \right) \frac{t^n}{[n]_{p,q}!} \right). \tag{37}$$

Therefore, we can see that

$$(p-q)pqx\mathcal{D}_{p,q}^{(2)}\mathcal{S}_{n,p,q}(x) = q\mathcal{D}_{p,q}\mathcal{S}_{n,p,q}(px) - p\mathcal{D}_{p,q}\mathcal{S}_{n,p,q}(qx), \tag{38}$$

and this shows the required result at once. □

Corollary 2.12. *From the Theorem 2.11, one holds*

$$\begin{aligned} & p^2qx\mathcal{D}_{p,q}^{(2)}\mathcal{S}_{n,p,q}(x) + p[n]_{p,q}\mathcal{S}_{n-1,p,q}(pqx) \\ & = q[n]_{p,q}\mathcal{S}_{n-1,p,q}(p^2x) + pq^2x\mathcal{D}_{p,q}^{(2)}\mathcal{S}_{n,p,q}(x). \end{aligned} \tag{39}$$

Theorem 2.13. *Let a, b be any positive integers. Then we find*

$$\sum_{k=0}^{n} \begin{bmatrix} n \\ k \end{bmatrix}_{p,q} \frac{\mathcal{S}_{n-k,p,q}(ax)\mathcal{S}_{k,p,q}(by)}{a^{n-k}b^k} = \sum_{k=0}^{n} \begin{bmatrix} n \\ k \end{bmatrix}_{p,q} \frac{\mathcal{S}_{n-k,p,q}(bx)\mathcal{S}_{k,p,q}(ay)}{a^k b^{n-k}}. \tag{40}$$

Proof. Suppose the form A is as the following.

$$A := \frac{e_{p,q}(tx)e_{p,q}(ty)}{\left(e_{p,q}\left(-\frac{t}{a}\right)+1\right)\left(e_{p,q}\left(-\frac{t}{b}\right)+1\right)}. \tag{41}$$

The form A of the equation above (41) can be transformed as

$$\begin{aligned} A &:= \sum_{n=0}^{\infty} \frac{\mathcal{S}_{n,p,q}(ax)t^n}{a^n [n]_{p,q}!} \sum_{n=0}^{\infty} \frac{\mathcal{S}_{n,p,q}(by)t^n}{b^n [n]_{p,q}!} \\ &= \sum_{n=0}^{\infty} \left(\sum_{k=0}^{n} \begin{bmatrix} n \\ k \end{bmatrix}_{p,q} \frac{\mathcal{S}_{n-k,p,q}(ax)\mathcal{S}_{k,p,q}(by)}{a^{n-k}b^k} \right) \frac{t^n}{[n]_{p,q}!}, \end{aligned} \tag{42}$$

or, equivalently,

$$A := \sum_{n=0}^{\infty} \left(\sum_{k=0}^{n} \begin{bmatrix} n \\ k \end{bmatrix}_{p,q} \frac{\mathcal{S}_{n-k,p,q}(bx)\mathcal{S}_{k,p,q}(ay)}{b^{n-k}a^k} \right) \frac{t^n}{[n]_{p,q}!}. \tag{43}$$

Comparing the coefficients of t^n in both sides, we find the required result. □

Corollary 2.14. *Setting $a = 1$ in the Theorem 2.13, one holds*

$$\sum_{k=0}^{n} \begin{bmatrix} n \\ k \end{bmatrix}_{p,q} \frac{\mathcal{S}_{n-k,p,q}(x)\mathcal{S}_{k,p,q}(by)}{b^k} = \sum_{k=0}^{n} \begin{bmatrix} n \\ k \end{bmatrix}_{p,q} \frac{\mathcal{S}_{n-k,p,q}(bx)\mathcal{S}_{k,p,q}(y)}{b^{n-k}}. \tag{44}$$

Corollary 2.15. *When $p = 1$ and $q \to 1$ in the Theorem 2.13, one holds*

$$\sum_{k=0}^{n} \binom{n}{k} \frac{S_{n-k}(ax)S_k(by)}{a^{n-k}b^k} = \sum_{k=0}^{n} \binom{n}{k} \frac{S_{n-k}(bx)lS_k(ay)}{a^k b^{n-k}}, \tag{45}$$

where $S_n(x)$ is the sigmoid polynomials (see [16]).

Theorem 2.16. *Let a, b be any integers without 0. Then we obtain*

$$\sum_{k=0}^{n}\begin{bmatrix} n \\ k \end{bmatrix}_{p,q}\frac{(-1)^n S_{n-k,p,q}(ax)E_{k,p,q}(by)}{a^{n-k}b^k}=\sum_{k=0}^{n}\begin{bmatrix} n \\ k \end{bmatrix}_{p,q}\frac{S_{n-k,p,q}(-bx)E_{k,p,q}(-ay)}{a^k b^{n-k}}, \quad (46)$$

where $E_{n,p,q}(x)$ is the (p, q)-Euler polynomials (see [10]).
Proof. We consider the form B as

$$B:=\frac{e_{p,q}(tx)e_{p,q}(ty)}{\left(e_{p,q}\left(-\frac{t}{a}\right)+1\right)\left(e_{p,q}\left(\frac{t}{b}\right)+1\right)}. \quad (47)$$

The form B of the equation above (47) can be transformed as

$$\begin{aligned} B:&=\frac{1}{[2]_{p,q}}\sum_{n=0}^{\infty}\frac{S_{n,p,q}(ax)t^n}{a^n[n]_{p,q}!}\sum_{n=0}^{\infty}\frac{E_{n,p,q}(by)t^n}{b^n[n]_{p,q}!} \\ &=\frac{1}{[2]_{p,q}}\sum_{n=0}^{\infty}\left(\sum_{k=0}^{n}\begin{bmatrix} n \\ k \end{bmatrix}_{p,q}\frac{S_{n-k,p,q}(ax)E_{k,p,q}(by)}{a^{n-k}b^k}\right)\frac{t^n}{[n]_{p,q}!}. \end{aligned} \quad (48)$$

Also, we can transform the form B such as

$$B:=\frac{1}{[2]_{p,q}}\sum_{n=0}^{\infty}\left(\sum_{k=0}^{n}\begin{bmatrix} n \\ k \end{bmatrix}_{p,q}\frac{S_{n-k,p,q}(-bx)E_{k,p,q}(-ay)}{(-1)^n b^{n-k}a^k}\right)\frac{t^n}{[n]_{p,q}!}. \quad (49)$$

Comparing the equation above (48) and (49), we derive the result of Theorem 2.16. □.

Corollary 2.17. *Setting $a = 1$ in the Theorem 2.16, one holds*

$$\sum_{k=0}^{n}\begin{bmatrix} n \\ k \end{bmatrix}_{p,q}\frac{(-1)^n S_{n-k,p,q}(x)E_{k,p,q}(by)}{b^k}=\sum_{k=0}^{n}\begin{bmatrix} n \\ k \end{bmatrix}_{p,q}\frac{S_{n-k,p,q}(-bx)E_{k,p,q}(-y)}{b^{n-k}}, \quad (50)$$

where $E_{n,p,q}(x)$ is the (p, q)-Euler polynomials.

Theorem 2.18. *Let a, b be any integers without 0. Then we have*

$$\sum_{k=0}^{n}\begin{bmatrix} n \\ k \end{bmatrix}_{p,q}\frac{S_{n-k,p,q}(-ax)E_{k,p,q}(by)}{a^{n-k}b^k}=\sum_{k=0}^{n}\begin{bmatrix} n \\ k \end{bmatrix}_{p,q}\frac{S_{n-k,p,q}(-bx)E_{k,p,q}(ay)}{a^k b^{n-k}}, \quad (51)$$

where $E_{n,p,q}(x)$ is the (p, q)-Euler polynomials.
Proof. We set the form C as

$$C:=\frac{e_{p,q}(tx)e_{p,q}(ty)}{\left(e_{p,q}\left(\frac{t}{a}\right)+1\right)\left(e_{p,q}\left(\frac{t}{b}\right)+1\right)}. \quad (52)$$

In the form C of the equation above (52), we can find that

$$\begin{aligned} C:&=\frac{1}{[2]_{p,q}}\sum_{n=0}^{\infty}\frac{S_{n,p,q}(-ax)t^n}{(-a)^n[n]_{p,q}!}\sum_{n=0}^{\infty}\frac{E_{n,p,q}(by)t^n}{b^n[n]_{p,q}!} \\ &=\frac{1}{[2]_{p,q}}\sum_{n=0}^{\infty}\left(\sum_{k=0}^{n}\begin{bmatrix} n \\ k \end{bmatrix}_{p,q}\frac{S_{n-k,p,q}(-ax)E_{k,p,q}(by)}{(-a)^{n-k}b^k}\right)\frac{t^n}{[n]_{p,q}!}, \end{aligned} \quad (53)$$

and

$$C:=\frac{1}{[2]_{p,q}}\sum_{n=0}^{\infty}\left(\sum_{k=0}^{n}\begin{bmatrix} n \\ k \end{bmatrix}_{p,q}\frac{\mathcal{S}_{n-k,p,q}(-bx)E_{k,p,q}(ay)}{(-b)^{n-k}a^{k}}\right)\frac{t^{n}}{[n]_{p,q}!}. \tag{54}$$

Comparing the both sides in the equation above (53) and (54), we complete the required result of Theorem 2.18.

Corollary 2.19. *Putting $a = -1$ in the Theorem 2.18, one holds*

$$\sum_{k=0}^{n}\begin{bmatrix} n \\ k \end{bmatrix}_{p,q}\frac{\mathcal{S}_{n-k,p,q}(x)E_{k,p,q}(by)}{b^{k}}=\sum_{k=0}^{n}\begin{bmatrix} n \\ k \end{bmatrix}_{p,q}\frac{\mathcal{S}_{n-k,p,q}(-bx)E_{k,p,q}(-y)}{(-1)^{n}b^{n-k}}, \tag{55}$$

where $E_{n,p,q}(x)$ is the (p,q)-Euler polynomials.

3. Structure and various phenomena of roots of $\mathcal{S}_{n,p,q}$ using the computer

This section mentions the structure of roots of $\mathcal{S}_{n,p,q}$. Furthermore, based on the previous content, an example of (p,q)-sigmoid polynomials is taken to identify the shape of the fixed points and the iterative function. And by applying it, we can find properties of self-similarity by using Newton's method.

$p = 0.9$	$p = 0.5$	$p = 0.2$
−1	−1	−0.99999
−0.390901 − 0.133409i	−0.319498 − 0.0466997i	−0.632357 − 0.0806528i
−0.390901 + 0.133409i	−0.319498 + 0.0466997i	−0.632357 + 0.0806528i
−0.325772 − 0.249143i	−0.22888 − 0.212259i	−0.574469 − 0.251005i
−0.325772 + 0.249143i	−0.22888 + 0.212259i	−0.574469 + 0.251005i
−0.229438 − 0.33589i	−0.149885 − 0.266689i	−0.458705 − 0.409513i
−0.229438 + 0.33589i	−0.149885 + 0.266689i	−0.458705 + 0.409513i
−0.113312 − 0.387358i	−0.0593879 − 0.293992i	−0.102928 − 0.574072i
−0.113312 + 0.387358i	−0.0593879 + 0.293992i	−0.102928 + 0.574072i
0.0110113 − 0.400647i	0.0334027 − 0.292274i	0.083275 − 0.559033i
0.0110113 + 0.400647i	0.0334027 + 0.292274i	0.083275 + 0.559033i
0.132056 − 0.37596i	0.119568 − 0.262599i	0.136077
0.132056 + 0.37596i	0.119568 + 0.262599i	0.24766 − 0.490886i
0.239081 − 0.316515i	0.18937	0.24766 + 0.490886i
0.239081 + 0.316515i	0.191091 − 0.208427i	0.379531 − 0.381018i
0.322803 − 0.228278i	0.191091 + 0.208427i	0.379531 + 0.381018i
0.322803 + 0.228278i	0.241247 − 0.134952i	0.471422 − 0.24088i
0.376059 − 0.119488i	0.241247 + 0.134952i	0.471422 + 0.24088i
0.376059 + 0.119488i	0.265157 − 0.0476321i	0.518528 − 0.0823508i
0.394325	0.265157 + 0.0476321i	0.518528 + 0.0823508i

Table 1.
Approximate zeros of $\mathcal{S}_{20,p,0.1}(x)$.

First, let us find an approximation of the root of (p, q)-sigmoid polynomials. At this time, p and q should not be the same value. If they have the same value, the denominator of (p, q)-sigmoid polynomials will be 0. Consider the roots of $\mathcal{S}_{20,p,q}$. Once we fix the value of q to 0.1 and switch p into 0.9, 0.5, and 0.2, the following **Table 1** can be found.

Here we can see that the approximate values of the roots change as the value of p changes, as the roots always contain two real roots. Therefore, we can make the following assumptions.

Conjecture 3.1. *Roots of* $\mathcal{S}_{20,p,q}$ *when* $|p| < 1$, $n = 20$ *and* $q = 0.1$, *always have two real roots.*

In the same way, we can find an approximation of the roots when the values of q are changed in order of 0.9, 0.5, and 0.2 and p is fixed at 0.1. In this case, when $p = 0.1$ and $q = 0.2$, the approximate roots of $\mathcal{S}_{20,p,q}$ all possess real roots which can be confirmed through Mathematica ($x = -395824, -136973, -47951, -16837, -5912, -2076, -728, -255, -89, -31, -10, -3, -1, -0, 2, 19, 157, 1231, 9610, 71387$).

Table 1 can be illustrated as **Figure 1**. The figure on the left is when $p = 0.9$ and the figure on the right is when $p = 0.2$. Here we can see that the structure of the roots is changing. The following will identify the structure and the build-up of the roots of $\mathcal{S}_{n,p,q}$. The structures of approximations roots in polynomials that combine the existing (p, q)-number can be found becoming closer to a circle as the n increases and can be seen that a single root is continuously stacked at a certain point. Also, as the roots continue to pile up near at some point and the larger the n becomes, it can be assumed that $\mathcal{S}_{20,p,q}$ becomes closer to the circle. Actually, a picture of $\mathcal{S}_{20,p,q}$ can be made using Mathematica.

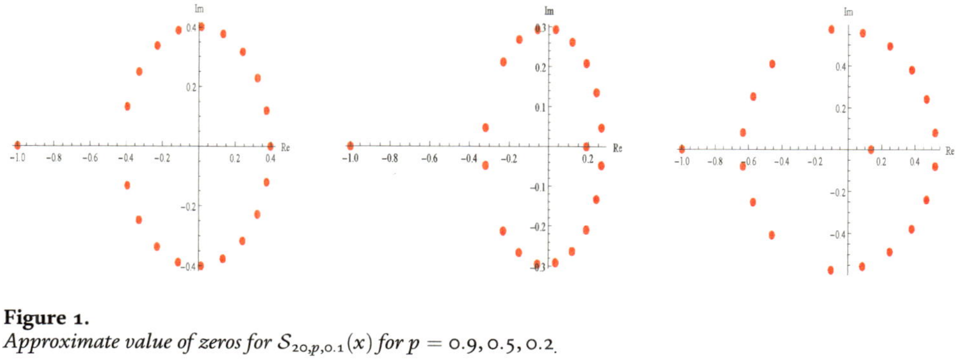

Figure 1.
Approximate value of zeros for $\mathcal{S}_{20,p,0.1}(x)$ *for* $p = 0.9, 0.5, 0.2$.

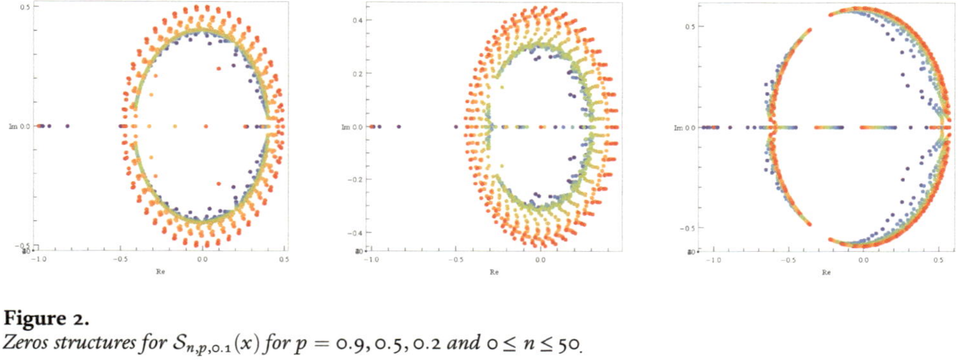

Figure 2.
Zeros structures for $\mathcal{S}_{n,p,0.1}(x)$ *for* $p = 0.9, 0.5, 0.2$ *and* $0 \leq n \leq 50$.

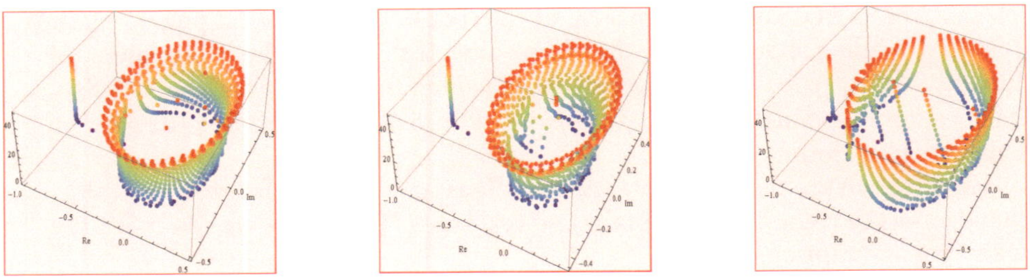

Figure 3.
Zeros scattering for $S_{n,p,0.1}(x)$ for $p = 0.9, 0.5, 0.2$ and $0 \leq n \leq 50$.

The following **Figure 2** shows the structure of the roots when n is 0 to 50. When fixed at $q = 0.1$, the figure on the left is when $p = 0.9$ and the right is when $p = 0.2$.

Three-dimensional identification of **Figure 2** shows the following **Figure 3**. Given the speculation, the roots are piling up near a point ($x = -1$) and as the p approaches 1, the rest of the roots become closer to a circle as the value of n increases.

Based on the content above, we will now look at fixed points of $S_{n,p,q}$. At this point, the value of q is fixed at 0.1 and p is changed to 0.9, 0.5, and 0.2 respectively. This can be found as shown in the following **Figure 4**. **Figure 4** shows a nearly circular appearance and always has the origin. Also, as the value of p decreases, it can be seen that the distance from the origin and the roots increases.

Similarly, the fixed points in the 3D structure can be checked as shown in **Figure 5**.

Conjecture 3.2. *$S_{n,p,q}$ may have one fixed point which is the origin and the rest of the fixed points appear in the form of a circle.*

The following is a third polynomial of $S_{3,p,q}$, using a iterative function to find the approximate value of the fixed point. First, by iterating this third function five

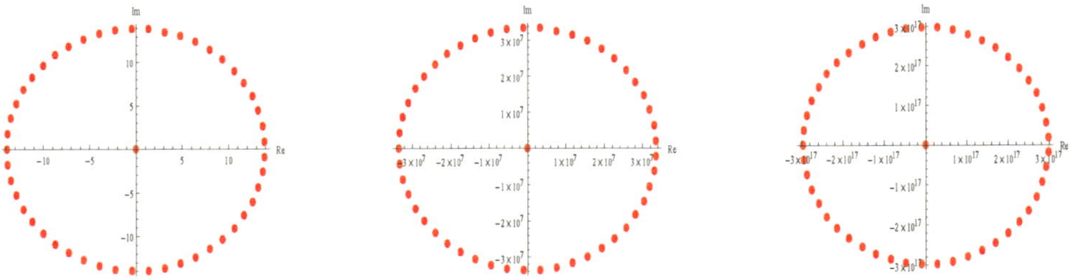

Figure 4.
Fixed points for $S_{50,p,0.1}(x)$ for $p = 0.9,\ 0.5,\ 0.2$.

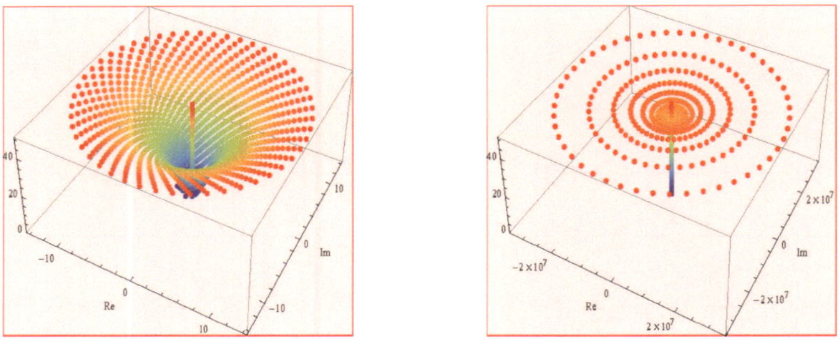

Figure 5.
Scattering of fixed points for $S_{n,p,0.1}(x)$ for $p = 0.9, 0.5$ and $0 \leq n \leq 50$.

times and getting the number of real roots, the value of fixed points will vary depending on the value of p. For $p = 0.9$ and $p = 0.5$, the number of the real roots of each of the five iterated function is 1, 1, 1, 1, 1, 1, but for $p = 0.2$, 3, 3, 3, 3, 3 appears. The structure of the approximate value of the actual fixed point is shown in **Figure 6**. The top part of **Figure 6** is when $p = 0.9$ and the bottom figures represent $p = 0.2$. Typically, most of the roots structures of general polynomials using q-numbers appear a circular shape, but it is difficult to find constant regularity in fixed points. However, sigmoid function including (p, q)-numbers might have a special property for fixed points. In other words, we can guess from **Figure 6** that the structure of fixed points for (p, q)-sigmoid polynomials will become a circular shape if n increases tremendously.

Let us look at the following by observing an application of an iterated $\mathcal{S}_{3,p,0.1}$ using **Figure 6**. Let us try using the Newton's method that we know well. Let us divide the values that go to the root of the tertiary function. First, fix p at 0.9 and limit the range of values of x and y from -4 to 4. Then the approximate root of $\mathcal{S}_{3,0.9,0.1}$ becomes -0.936067, $0.187169 - 0.256352i$, $0.187169 + 0.256352i$. Also, if the values going to -0.93606 are shown in red, $0.187169 - 0.256352i$ in blue, $0.187169 + 0.256352i$ in yellow, it is shown as the left of **Figure 7**. The right side of **Figure 7** is the picture that comes when $\mathcal{S}_{3,0.9,0.1}$ are iterated twice.

The structure of the roots of $\mathcal{S}_{n,p,q}$ appears to have one value near -1 and become a circular form as the n increases. Also, as the value of p increases, the

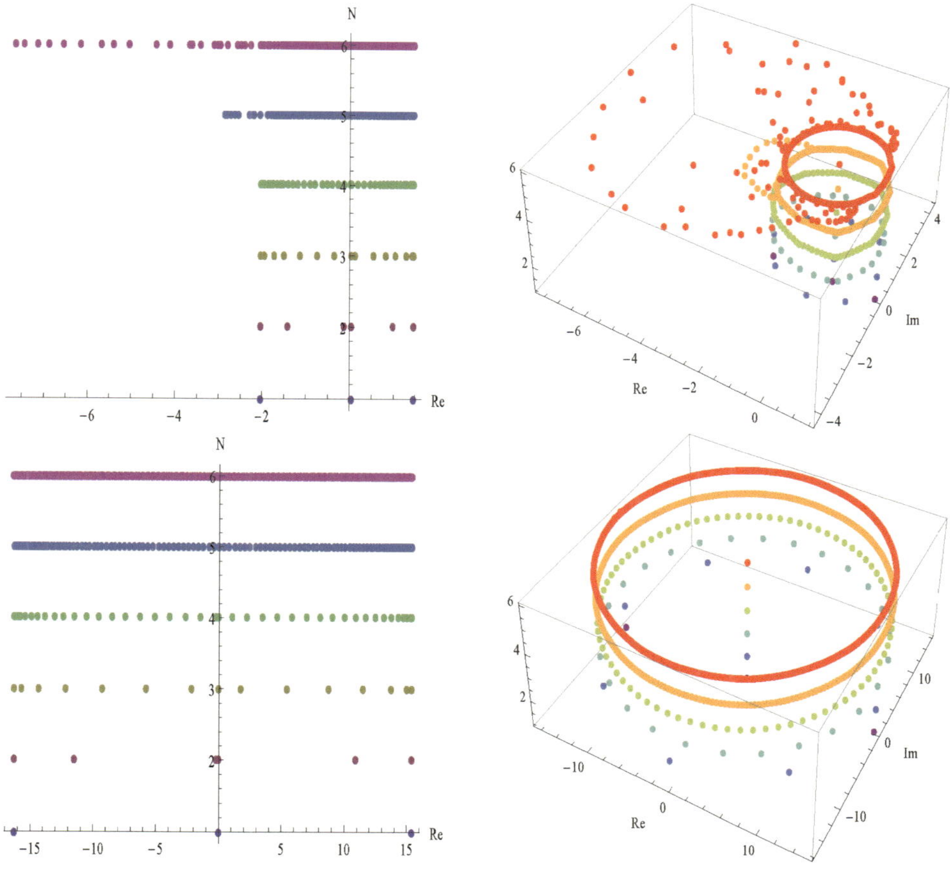

Figure 6.
Scattering of fixed points for $\mathcal{S}_{3,p,0.1}(x)$ iterated 5-times.

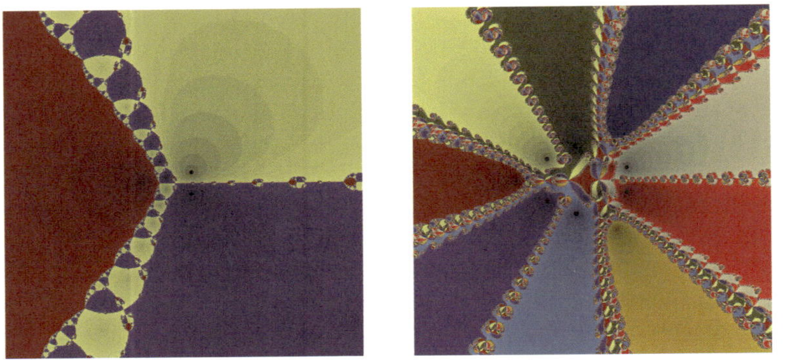

Figure 7.
Classification of values that go near the fixed points of $\mathcal{S}_{3,0.9,0.1}$.

diameter of the circle increases. A fixed point of $\mathcal{S}_{n,p,q}$ can be seen to have a nearly constant form as p is reduced, which can also confirm an increase in the radius.

4. Conclusion

Sigmoid function is a very important function in deep learning. In the current situation of artificial intelligence development, the properties and speculations of the sigmoid polynomials revealed in this paper in the area of using (p, q)-number could be an useful data in deep learning using activation functions. Through iterating $\mathcal{S}_{n,p,q}$ with these properties, it can be assumed to have self-similarity and can be studied further to confirm new properties.

Acknowledgements

This research was supported by Basic Science Research Program through the National Research Foundation of Korea (NRF) funded by the Ministry of Science, ICT and Future Planning (No. 2017R1E1A1A03070483).

Additional information

Mathematics Subject Classification: 11B68, 11B75, 12D10

Author details

Jung Yoog Kang
Department of Mathematics Education, Silla University, Busan, Republic of Korea

*Address all correspondence to: jykang@silla.ac.kr

References

[1] Chakrabarti R, Jagannathan R. A (p,q)-oscillator realization of two-parameter quantum algebras. Journal of Physics A: Mathematical and General. 1991;**24**:L711

[2] Arik M, Demircan E, Turgut T, Ekinci L, Mungan M. Fibonacci oscillators. Zeitschrift für Physik C Particles and Fields. 1991;**55**:89-95

[3] Brodimas G, Jannussis A, Mignani R. Two-Parameter Quantum Groups. Universita di Roma; Preprint, Nr. 820; 1991. Available from: https://inspirehep.net/literature/319903

[4] Wachs M, White D. (p,q)-Stirling numbers and set partition statistics. Journal of Combinatorial Theory, Series A. 1991;**56**:27-46

[5] Andrews GE, Askey R, Roy R. Special Functions. Cambridge, UK: Cambridge Press; 1999

[6] Araci S, Duran U, Acikgoz M, Srivastava HM. A certain (p,q)-derivative operator and associated divided differences. Journal of Inequalities and Applications. 2016; **301**:1-8

[7] Burban M, Klimyk AU. (P,Q)-differentiation, (P,Q)-integration and (P,Q)-hypergeometric functions related to quantum groups. Integral Transforms and Special Functions. 1994;**2**:15-36

[8] Cieslinski JL. Improved q-exponential and q-trigonometric functions. 2010. arXiv:1006.5652v1 [math.CA]

[9] Corcino RB. On (P,Q)-Binomial coefficients. Integers Electronic Journal of Combinatorial Number Theory. 2008;**8**:A29

[10] Duran U, Acikgoz M, Araci S. On (p,q)-Bernoulli, (p,q)-Euler and (p,q)-Genocchi polynomials. Journal of Computational and Theoretical Nanoscience. 2016

[11] Han J, Moraga C. The influence of the sigmoid function parameters on the speed of backpropagation learning. International Conference on Artificial Neural Networks. 2005;**930**:195-201

[12] Han J, Wilson RS, Leurgans SE. Sigmoidal mixed models for longitudinal data. Statistical Methods in Medical Research. March 2018;**27**(3):863-875. DOI: 10.1177/0962280216645632

[13] Jagannathan R, Rao KS. Two-parameter quantum algebras, twin-basic numbers, and associated generalized hypergeometric series. In: Proceeding of the International Conference on Number Theory and Mathematical Physics; 20–21 December 2005. Kumbakonam, India: Srinivasa Ramanujan Centre; 2005

[14] Jagannathan R. (P,Q)-special functions, special functions and differential equations. In: Proceedings of a Workshop Held at the Institute of Mathematical Sciences, Madras, India, January. 1997. pp. 13-24

[15] Kwan HK. Simple sigmoid-like activation function suitable for digital hardware implementation. Statistical Methods in Medical Research. 1992;**28**: 1379-1380. DOI: 10.1049/el:19920877

[16] Kang JY. Some relationships between sigmoid polynomials and other polynomials. Journal of Pure and Applied Mathematics. 2019;**1**(1–2):57-67

[17] Kang JY, Ryoo CS. A numerical investigation on the structure of the zeros of the q-tangent polynomials. Polynomials - Theory and Application. 2019:1-18. Available from: https://www.intechopen.com/books/polynomials-theory-and-application

[18] Ryoo CS, Kang JY. A numerical investigation on the structure of the zeros of the Euler polynomials. Discrete Dynamics in Nature and Society. 2015: 1-9. DOI: 10.1155/2015/174173

[19] Ryoo CS. A numerical investigation on the zeros of the tangent polynomials. Journal of Applied Mathematics and Informatics. 2014;**32**:315-322

[20] Kurt B. Relations on the Apostol type (p, q)-Frobenius-Euler polynomials and generalizations of the Srivastava-Pinter addition theorems. Turkish Journal of Analysis and Number Theory. 2017;**5**:126131

[21] Sadjang PN. On the fundamental theorem of (p, q)-calculus and some (p, q)-Taylor formulas. 2013. arXiv: 1309.3934 [math.QA]

Chapter 8

$\mathcal{I}$–Convergence of Arithmetical Functions

Vladimír Baláž and Tomáš Visnyai

Abstract

Let $n > 1$ be an integer with its canonical representation, $n = p_1^{\alpha_1} p_2^{\alpha_2} \cdots p_k^{\alpha_k}$. Put $H(n) = \max\{\alpha_1, \ldots, \alpha_k\}$, $h(n) = \min\{\alpha_1, \ldots, \alpha_k\}$, $\omega(n) = k$, $\Omega(n) = \alpha_1 + \cdots + \alpha_k$, $f(n) = \prod_{d|n} d$ and $f^*(n) = \frac{f(n)}{n}$. Many authors deal with the statistical convergence of these arithmetical functions. For instance, the notion of normal order is defined by means of statistical convergence. The statistical convergence is equivalent with $\mathcal{I}_d$–convergence, where $\mathcal{I}_d$ is the ideal of all subsets of positive integers having the asymptotic density zero. In this part, we will study $\mathcal{I}$–convergence of the well-known arithmetical functions, where $\mathcal{I} = \mathcal{I}_c^{(q)} = \left\{A \subset \mathbb{N} : \sum_{a \in A} a^{-q} < +\infty\right\}$ is an admissible ideal on $\mathbb{N}$ such that for $q \in (0, 1\rangle$ we have $\mathcal{I}_c^{(q)} \subsetneq \mathcal{I}_d$, thus $\mathcal{I}_c^{(q)}$–convergence is stronger than the statistical convergence ($\mathcal{I}_d$–convergence).

Keywords: sequences, $\mathcal{I}$–convergence, arithmetical functions, normal order, binomial coefficients

1. Introduction

The notion of statistical convergence was introduced independently by Fast and Schoenberg in [1, 2], and the notion of $\mathcal{I}$–convergence introduced by Kostyrko et al. in the paper [3] coresponds to the natural generalization of statistical convergence (see also [4] where $\mathcal{I}$–convergence is defined by means of filter – the dual notion to ideal). These notions have been developed in several directions in [5–18] and have been used in various parts of mathematics, in particular in Number Theory and Ergodic Theory, for example [15, 19–28] also in Economic Theory [29, 30] and Political Science [31]. Many authors deal with average and normal order of the well-known arithmetical functions (see [20, 21, 23, 24, 26, 28, 32, 33] and the monograph [34] for basic properties of the well-known arithmetical functions). In what follows, we shall strengthen these results from the standpoint of $\mathcal{I}$–convergence of sequences, mainly by $\mathcal{I}_c^{(q)}$–convergence and $\mathcal{I}_u$–convergence. On connection with that we can obtain a good information about behaviour and properties of the well-known arithmetical functions by investigating $\mathcal{I}$–convergence of these functions or some sequences connected with these functions. Specifically in [28] by means of $\mathcal{I}_d$–convergence, there is recalled the result that normal order of $\Omega(n)$ or $\omega(n)$ respectively is $\log \log n$. We managed to completely determine for which $q \in (0, 1\rangle$ the sequences $\frac{\Omega(n)}{\log \log n}$ and $\frac{\omega(n)}{\log \log n}$ are $\mathcal{I}_c^{(q)}$–convergent. As consequence of our results, we have that the above sequences are $\mathcal{I}_d$–convergent to 1, what is equivalent that normal order of $\Omega(n)$ or $\omega(n)$ respectively is $\log\log n$. Further in [26], there is proved that the sequence $\left(\log p \cdot \frac{a_p(n)}{\log n}\right)$ is $\mathcal{I}_d$–convergent to 0 (see also [21]). We shall extend this result by means of $\mathcal{I}_u$–convergence of the sequence $\left(\log p \cdot \frac{a_p(n)}{\log n}\right)$. So we can get a better view of the structure of the set $B(\varepsilon) = \left\{n \in \mathbb{N} : \log p \cdot \frac{a_p(n)}{\log n} < \varepsilon\right\}$, $\varepsilon > 0$. We also want to investigate the $\mathcal{I}_c^{(q)}$–convergence of further arithmetical functions.

2. Basic notions

Let $\mathbb{N}$ be the set of positive integers. Let $A \subseteq \mathbb{N}$. If $m, n \in \mathbb{N}$, $m \leq n$, we denote by $A(m,n)$ the cardinality of the set $A \cap [m,n]$. $A(1,n)$ is abbreviated by $A(n)$. We recall the concept of asymthotic, logarithmic and uniform density of the set $A \subseteq \mathbb{N}$ (see [35–38]).

Definition 1.1. *Let* $A \subseteq \mathbb{N}$.

a. Put $d_n(A) = \frac{A(n)}{n} = \frac{1}{n}\sum_{k=1}^{n}\chi_A(k)$, where χ_A is the characteristic function of the set A. Then the numbers $\underline{d}(A) = \liminf_{n\to\infty} d_n(A)$ and $\overline{d}(A) = \limsup_{n\to\infty} d_n(A)$ are called the *lower* and *upper asymptotic density* of the set A, respectively. If there exists $\lim_{n\to\infty} d_n(A)$, then $d(A) = \underline{d}(A) = \overline{d}(A)$ is said to be the *asymptotic density* of A.

b. Put $\delta_n(A) = \frac{1}{s_n}\sum_{k=1}^{n}\frac{\chi_A(k)}{k}$, where $s_n = \sum_{k=1}^{n}\frac{1}{k}$. Then the numbers $\underline{\delta}(A) = \liminf_{n\to\infty}\delta_n(A)$ and $\overline{\delta}(A) = \limsup_{n\to\infty}\delta_n(A)$ are called the *lower* and *upper logarithmic density* of A, respectively. Similarly, if there exists $\lim_{n\to\infty}\delta_n(A)$, then $\delta(A) = \underline{\delta}(A) = \overline{\delta}(A)$ is said to be the *logarithmic density* of A. Since $s_n = \log n + \gamma + O\left(\frac{1}{k}\right)$ for $n \to \infty$ and γ is the Euler constant, s_n can be replaced by $\log n$ in the definition of $\delta_n(A)$.

c. Put $\alpha_s = \min_{n\geq 0} A(n+1, n+s)$ and $\alpha^s = \max_{n\geq 0} A(n+1, n+s)$. The following limits $\underline{u}(A) = \lim_{s\to\infty}\frac{\alpha_s}{s}$, $\overline{u}(A) = \lim_{s\to\infty}\frac{\alpha^s}{s}$ exist (see [17, 37, 39, 40]) and they are called *lower* and *upper uniform density* of the set A, respectively. If $\overline{u}(A) = \underline{u}(A)$, then we denote it by $u(A)$ and it is called the *uniform density* of A. It is clear that for each $A \subseteq \mathbb{N}$ we have

$$\underline{u}(A) \leq \underline{d}(A) \leq \underline{\delta}(A) \leq \overline{\delta}(A) \leq \overline{d}(A) \leq \overline{u}(A). \tag{1}$$

Further densities can be found in papers [11, 12].

Let $n = p_1^{\alpha_1}p_2^{\alpha_2}\cdots p_k^{\alpha_k}$ be the canonical representation of the integer $n \in \mathbb{N}$. Recall some arithmetical functions, which belong to our interest.

1. $\omega(n)$ – the number of distinct prime factors of n ($\omega(n) = k$),

2. $\Omega(n)$ – the number of prime factors of n counted with multiplicities ($\Omega(n) = \alpha_1 + \cdots + \alpha_k$),

3. $d(n)$ – the number of divisors of n $\left(d(n) = \sum_{d|n} 1\right)$,

DOI: http://dx.doi.org/10.5772/intechopen.91932

4. define $h(n)$ and $H(n)$, put $h(1) = 1$, $H(1) = 1$ and for $n > 1$ denote

$$h(n) = \min_{1 \le j \le k} \alpha_j, \quad H(n) = \max_{1 \le j \le k} \alpha_j,$$

5. $f(n) = \prod_{d|n} d$, $f^*(n) = \frac{f(n)}{n}$, where $n = 1, 2, \dots,$

6. let p be a prime number, $a_p(n)$ is defined as follows: $a_p(1) = 0$ and if $n > 0$, then $a_p(n)$ is a unique integer $j \ge 0$ satisfying $p^j | n$, but $p^{j+1} \nmid n$ i.e., $p^{a_p(n)} \| n$,

7. $\gamma(n)$ and $\tau(n)$ – were introduced in connection with representation of natural numbers of the form $n = a^b$, where a, b are positive integers. Let

$$n = a_1^{b_1} = a_2^{b_2} = \dots = a_{\gamma(n)}^{b_{\gamma(n)}}$$

be all such representations of given natural number n, where $a_i, b_i \in \mathbb{N}$. Denote by

$$\tau(n) = b_1 + b_2 + \dots + b_{\gamma(n)}, \quad (n > 1).$$

It is clear that $\gamma(n) \ge 1$, because for any $n > 1$ there exist representation in the form n^1.

3. Ideals

A lot of mathematical disciplines use the term small (large) set from different point of view. For instance a final set, a set having the measure zero and nowhere dense set is a small set from point of view of cardinality, measure (probability) and topology, respectively. The notion of ideal $\mathcal{I} \subseteq 2^X$ is the unifying principle how to express that a subset of $X \neq \emptyset$ is small. We say a set $A \subseteq X$ is a small set if $A \in \mathcal{I}$. Recall the notion of an ideal $\mathcal{I}$ of subsets of $\mathbb{N}$.

Let $\mathcal{I} \subseteq 2^{\mathbb{N}}$. $\mathcal{I}$ is said to be an *ideal* in $\mathbb{N}$, if $\mathcal{I}$ is additive (if $A, B \in \mathcal{I}$ then $A \cup B \in \mathcal{I}$) and hereditary (if $A \in \mathcal{I}$ and $B \subset A$ then $B \in \mathcal{I}$). An ideal $\mathcal{I}$ is said to be *non-trivial ideal* if $\mathcal{I} \neq \emptyset$ and $\mathbb{N} \notin \mathcal{I}$. A non-trivial ideal $\mathcal{I}$ is said to be *admissible ideal* if it contains all finite subsets of $\mathbb{N}$. The dual notion to the ideal is the notion filter. A non-empty family of sets $\mathcal{F} \subset 2^{\mathbb{N}}$ is a *filter* if and only if $\emptyset \notin \mathcal{F}$, for each $A, B \in \mathcal{F}$ we have $A \cap B \in \mathcal{F}$ and for each $A \in \mathcal{F}$ and each $B \supset A$ we have $B \in \mathcal{F}$ (for definitions see e.g. [4, 41, 42]). Let $\mathcal{I}$ be a proper ideal in $\mathbb{N}$ (i.e. $\mathbb{N} \notin \mathcal{I}$). Then a family of sets $\mathcal{F}(\mathcal{I}) = \{B \subseteq \mathbb{N} : \textit{there exists } A \in \mathcal{I} \textit{ such that } B = \mathbb{N} \setminus A\}$ is a filter in $\mathbb{N}$, so called the *associated filter* with the ideal $\mathcal{I}$.

The following example shows the most commonly used admissible ideals in different areas of mathematics.

Example 1.2.

a. The class of all finite subsets of $\mathbb{N}$ forms an admissible ideal usually denoted by $\mathcal{I}_f$.

b. Let ϱ be a density function on $\mathbb{N}$, the set $\mathcal{I}_\varrho = \{A \subset \mathbb{N} : \varrho(A) = 0\}$ is an admissible ideal. We will use namely the ideals $\mathcal{I}_d$, $\mathcal{I}_\delta$ and $\mathcal{I}_u$ related to asymptotic, logarithmic and uniform density, respectively.

c. A wide class of ideals $\mathcal{I}$ can be obtained by means of regular non negative matrixes $\mathbf{T} = (t_{n,k})_{n,k \in \mathbb{N}}$ (see [43]). For $A \subset \mathbb{N}$, we put $d_{\mathbf{T}}^{(n)}(A) = \sum_{k=1}^{\infty} t_{n,k} \chi_A(k)$

for $n \in \mathbb{N}$. If $\lim_{n\to\infty} d_{\mathbf{T}}^{(n)}(A) = d_{\mathbf{T}}(A)$ exists, then $d_{\mathbf{T}}(A)$ is called **T**–*density* of A (see [3, 44]). Put $\mathcal{I}_{d_{\mathrm{T}}} = \{A \subset \mathbb{N} : d_{\mathrm{T}}(A) = 0\}$. Then $\mathcal{I}_{d_{\mathrm{T}}}$ is a non-trivial ideal and $\mathcal{I}_{d_{\mathrm{T}}}$ contains both $\mathcal{I}_d$ and $\mathcal{I}_\delta$ ideals as a special case. Indeed $\mathcal{I}_d$ can be obtained by choosing $t_{n,k} = \frac{1}{n}$ for $k \le n$, $t_{n,k} = 0$ for $k > n$ and $\mathcal{I}_\delta$ by choosing $t_{n,k} = \frac{\frac{1}{k}}{s_n}$ for $k \le n$, $t_{n,k} = 0$ for $k > n$ where $s_n = \sum_{k=1}^{n} \frac{1}{k}$ for $n \in \mathbb{N}$.

For the matrix $\mathbf{T} = (t_{n,k})_{n,k\in\mathbb{N}}$, where $t_{n,k} = \frac{\varphi(k)}{n}$ for $k \le n$, $k|n$ and $t_{n,k} = 0$ otherwise we obtain $\mathcal{I}_\varphi$ ideal of Schoenberg (see [2]), where φ is Euler function.

Another special case of $\mathcal{I}_{d_{\mathrm{T}}}$ is the following. Take an arbitrary divergent series $\sum_{n=1}^{\infty} c_n$, where $c_n > 0$ for $n \in \mathbb{N}$ and put $t_{n,k} = \frac{c_k}{S_n}$ for $k \le n$, where $S_n = \sum_{i=1}^{n} c_i$, and $t_{n,k} = 0$ for $k > n$.

d. Let μ be a finitely additive normed measure on a field $\mathcal{S} \subseteq 2^{\mathbb{N}}$. Suppose that $\mathcal{S}$ contains all singletons $\{n\}$, $n \in \mathbb{N}$. Then the family $\mathcal{I}_\mu = \{A \subset \mathbb{N} : \mu(A) = 0\}$ is an admissible ideal. In the case if μ is the Buck measure density (see [13, 45]), $\mathcal{I}_\mu$ is an admissible ideal and $\mathcal{I}_\mu \subsetneq \mathcal{I}_d$.

e. Suppose that $\mu_n : 2^{\mathbb{N}} \to [0, 1]$ is a finitely additive normed measure for $n \in \mathbb{N}$. If for $A \subseteq \mathbb{N}$ there exists $\mu(A) = \lim_{n\to\infty} \mu_n(A)$, then the set A is said to be *measurable* and $\mu(A)$ is called the *measure* of A. Obviously μ is a finitely additive measure on some field $\mathcal{S} \subseteq 2^{\mathbb{N}}$. The family $\mathcal{I}_\mu = \{A \subset \mathbb{N} : \mu(A) = 0\}$ is a non-trivial ideal. For μ_n we can take for instance d_n, δ_n or $d_{\mathbf{T}}^{(n)}$.

f. Let $\mathbb{N} = \bigcup_{j=1}^{\infty} D_j$ be a decomposition on $\mathbb{N}$ (i.e. $D_k \cap D_l = \emptyset$ for $k \ne l$). Assume that D_j ($j = 1, 2, \ldots$) are infinite sets (e.g. we can choose $D_j = \{2^{j-1} \cdot (2s-1) : s \in \mathbb{N}\}$ for $j = 1, 2, \ldots$). Denote $\mathcal{I}_{\mathbb{N}}$ the class of all $A \subset \mathbb{N}$ such that A intersects only a finite number of D_j. Then $\mathcal{I}_{\mathbb{N}}$ is an admissible ideal.

g. For an $q \in (0, 1\rangle$ the set $\mathcal{I}_c^{(q)} = \{A \subset \mathbb{N} : \sum_{a\in A} a^{-q} < +\infty\}$ is an admissible ideal (see [23]). The ideal $\mathcal{I}_c^{(1)} = \{A \subset \mathbb{N} : \sum_{a\in A} a^{-1} < +\infty\}$ is usually denoted by $\mathcal{I}_c$. It is easy to see, that for any $q_1, q_2 \in (0, 1)$, $q_1 < q_2$ we have

$$\mathcal{I}_f \subsetneq \mathcal{I}_c^{(q_1)} \subsetneq \mathcal{I}_c^{(q_2)} \subsetneq \mathcal{I}_c \subsetneq \mathcal{I}_d \subsetneq \mathcal{I}_\delta. \tag{2}$$

The fact $\mathcal{I}_c \subsetneq \mathcal{I}_d$ in Eq. (2) follows from the following result. Let $A \subseteq \mathbb{N}$ and $\sum_{a\in A} \frac{1}{a} < \infty$ then $d(A) = 0$ (see [46]) thus if $A \in \mathcal{I}_c$ then $A \in \mathcal{I}_d$. The opposite is not true, consider the set of primes $\mathbb{P}$, for which we have $d(\mathbb{P}) = 0$ but $\sum_{p\in\mathbb{P}} \frac{1}{p} = \infty$ thus $\mathbb{P} \in \mathcal{I}_d$ but $\mathbb{P} \notin \mathcal{I}_c$ ($\mathcal{I}_c \ne \mathcal{I}_d$).

The fact that for any $q_1, q_2 \in (0, 1\rangle$, $q_1 < q_2$ we have $\mathcal{I}_c^{(q_1)} \subsetneq \mathcal{I}_c^{(q_2)}$ in Eq. (2) is clear. For showing that $\mathcal{I}_c^{(q_1)} \ne \mathcal{I}_c^{(q_2)}$ it suffices to find a set $H = \{h_1 < h_2 < \cdots < h_k < \cdots\} \subset \mathbb{N}$ such that $\sum_{k=1}^{\infty} h_k^{-q_1} = +\infty$ and $\sum_{k=1}^{\infty} h_k^{-q_2} < +\infty$. Put $h_k = \left[k^{\frac{1}{q_1}}\right]$. Since $h_1 < h_2 < \cdots < h_k < \cdots$ and $h_k^{q_1} \le k$ we have $\sum_{k=1}^{\infty} h_k^{-q_1} \ge \sum_{k=1}^{\infty} k^{-1} = +\infty$. On the other side $h_k > k^{\frac{1}{q_1}} - 1 \ge \frac{1}{2} k^{\frac{1}{q_1}}$ for $k \ge 2$, so we obtain $\sum_{k=1}^{\infty} h_k^{-q_2} \le 2^{q_2} \sum_{k=1}^{\infty} k^{-\frac{q_2}{q_1}} < +\infty$ since $\frac{q_2}{q_1} > 1$.

4. $\mathcal{I}$– and $\mathcal{I}^*$–convergence

The notion of statistical convergence was introduced in [1, 2] and the notion of $\mathcal{I}$–convergence introduced in [3] corresponds to the natural generalization of the notion of statistical convergence.

Let us recall notions of statistical convergence, $\mathcal{I}$– and $\mathcal{I}^*$–convergence of sequence of real numbers (see [3]).

Definition 1.3. We say that a sequence $(x_n)_{n=1}^{\infty}$ is *statistically convergent* to a number $L \in \mathbb{R}$ and we write $\lim \text{stat} x_n = L$, provided that for each $\varepsilon > 0$ we have $d(A(\varepsilon)) = 0$, where $A(\varepsilon) = \{n \in \mathbb{N} : |x_n - L| \geq \varepsilon\}$.

Definition 1.4.

i. We say that a sequence $(x_n)_{n=1}^{\infty}$ is *$\mathcal{I}$–convergent* to a number $L \in \mathbb{R}$ and we write $\mathcal{I} - \lim x_n = L$, if for each $\varepsilon > 0$ the set $A(\varepsilon) = \{n \in \mathbb{N} : |x_n - L| \geq \varepsilon\}$ belongs to the ideal $\mathcal{I}$.

ii. Let $\mathcal{I}$ be an admissible ideal on $\mathbb{N}$. A sequence $(x_n)_{n=1}^{\infty}$ of real numbers is said to be *$\mathcal{I}^*$–convergent* to $L \in \mathbb{R}$, if there is a set $H \in \mathcal{I}$, such that for $M = \mathbb{N} \backslash H = \{m_1 < m_2 < \cdots < m_k < \cdots\} \in \mathcal{F}(\mathcal{I})$ we have $\lim_{k \to \infty} x_{m_k} = L$, where the limit is in the usual sense.

In the definition of usual convergence the set $A(\varepsilon)$ is finite, it means that it is small from point of view of cardinality, $A(\varepsilon) \in \mathcal{I}_f$. Similarly in the definition of statistical convergence the set $A(\varepsilon)$ has asymptotic density zero, it is small from point of view of density, $A(\varepsilon) \in \mathcal{I}_d$. The natural generalization of these notions is the following, let $\mathcal{I}$ be an admissible ideal (e.g. anyone from Example 1.2) then for each $\varepsilon > 0$ we ask whether the set $A(\varepsilon)$ belongs in the ideal $\mathcal{I}$. In this way we obtain the notion of the $\mathcal{I}$–convergence. For the following use, we note that the concept of $\mathcal{I}$–convergence can be extended for such sequences that are not defined for all $n \in \mathbb{N}$, but only for "almost" all $n \in \mathbb{N}$. This means that instead of a sequence $(x_n)_{n=1}^{\infty}$ we have $(x_s)_{s \in S}$, where s runs over all positive integers belonging to $S \subseteq \mathbb{N}$ and $S \in \mathcal{F}(\mathcal{I})$.

Remember that $\mathcal{I}$–convergence in $\mathbb{R}$ has many properties similar to properties of the usual convergence. All notions which are used next we considered in real numbers $\mathbb{R}$. The following theorem can be easily proved.

Theorem 1.5 (Theorem 2.1 from [9]).

i. If $\mathcal{I} - \lim x_n = L$ and $\mathcal{I} - \lim y_n = K$, then $\mathcal{I} - \lim (x_n \pm y_n) = L \pm K$.

ii. If $\mathcal{I} - \lim x_n = L$ and $\mathcal{I} - \lim y_n = K$, then $\mathcal{I} - \lim (x_n \cdot y_n) = L \cdot K$.

The following properties are the most familiar axioms of convergence (see [47]).

(S) Every constant sequence $(x, x, \ldots, x, \ldots)$ converges to x.

(H) The limit of any convergent sequence is uniquely determined.

(F) If a sequence $(x_n)_{n=1}^{\infty}$ has the limit L, then each of its subsequences has the same limit.

(U) If each subsequence of the sequence $(x_n)_{n=1}^{\infty}$ has a subsequence which converges to L, then $(x_n)_{n=1}^{\infty}$ converges to L.

A natural question arises which above axioms are satisfied for the concept of $\mathcal{I}$–convergence.

Theorem 1.6 (see [14] and Proposition 3.1 from [3], where the concept of $\mathcal{I}$–convergence has been investigated in a metric space) *Let* $\mathcal{I} \subset 2^{\mathbb{N}}$ *be an admissible ideal.*

i. $\mathcal{I}$–convergence satisfies (S), (H) and (U).

ii. If $\mathcal{I}$ contains an infinite set, then $\mathcal{I}$–convergence does not satisfy (F).

Theorem 1.7 (see [3]) *Let* $\mathcal{I}$ *be an admissible ideal in* $\mathbb{N}$. *If* $\mathcal{I}^* - \lim x_n = L$ *then* $\mathcal{I} - \lim x_n = L$.

The following example shows that the converse of Theorem 1.7 is not true.

Example 1.8. Let $\mathcal{I} = \mathcal{I}_{\mathbb{N}}$ be an ideal from Example 1.2 f). Define $(x_n)_{n=1}^{\infty}$ as follows: For $n \in D_j$ we put $x_n = \frac{1}{j}$ for $j = 1, 2, \dots$. Then obviously $\mathcal{I} - \lim x_n = 0$. But we show that $\mathcal{I}^* - \lim x_n = 0$ does not hold.

If $H \in \mathcal{I}$ then directly from the definition of $\mathcal{I}$ there exists $p \in \mathbb{N}$ such that $H \subseteq D_1 \cup D_2 \cup \cdots \cup D_p$. But then $D_{p+1} \subseteq \mathbb{N} \backslash H = \{m_1 < m_2 < \cdots < m_k < \cdots\} \in \mathcal{F}(\mathcal{I})$ and so we have $x_{m_k} = \frac{1}{p+1}$ for infinitely many indices $k \in \mathbb{N}$. Therefore $\lim_{k \to \infty} x_{m_k} = 0$ cannot be true.

In [3] was formulated a necessary and sufficient condition for an admissible ideal $\mathcal{I}$ under which $\mathcal{I}$– and $\mathcal{I}^*$–convergence to be equivalent. Recall this condition (AP) that is similar to the condition (APO) in [7, 35].

Definition 1.9 (see also [40]) An admissible ideal $\mathcal{I} \subset 2^{\mathbb{N}}$ is said to satisfy the condition (AP) if for every countable family of mutually disjoint sets $\{A_1, A_2, \dots\}$ belonging to $\mathcal{I}$ there exists a countable family of sets $\{B_1, B_2, \dots\}$ such that symmetric difference $A_j \Delta B_j$ is finite for $j \in \mathbb{N}$ and $B = \cup_{j=1}^{\infty} B_j \in \mathcal{I}$.

Remark. Observe that each B_j from the previous Definition belong to $\mathcal{I}$.

Theorem 1.10 (see [14]) *From* $\mathcal{I} - \lim x_n = L$ *the statement* $\mathcal{I}^* - \lim x_n = L$ *follows if and only if* $\mathcal{I}$ *satisfies the condition (AP).*

In [44] it is proved that $\mathcal{I}_{d_{\mathrm{T}}}$– and $\mathcal{I}^*_{d_{\mathrm{T}}}$–convergence are equivalent in $\mathbb{R}$ provided that $\mathbf{T} = (t_{n,k})_{n,k \in \mathbb{N}}$ from Example 1.2 c) is a non-negative triangular matrix with $\sum_{k=1}^{n} t_{n,k} = 1$ for $n \in \mathbb{N}$. From this we get that $\mathcal{I}_d, \mathcal{I}_\delta, \mathcal{I}_\varphi$–convergence coinside with $\mathcal{I}^*_d, \mathcal{I}^*_\delta, \mathcal{I}^*_\varphi$–convergence, respectively. On the other hand for further ideals from Example 1.2 e.g. $\mathcal{I}_u$, $\mathcal{I}_{\mathbb{N}}$ and $\mathcal{I}_\mu$, respectively, we have that they do not fulfill the assertion that their $\mathcal{I}$–convergence coincides with $\mathcal{I}^*$–convergence. Since these ideals do not fulfill condition (AP) (see [13, 38, 40]).

The following Theorem shows that also for all ideals $\mathcal{I}_c^{(q)}$ for $q \in (0, 1\rangle$ the concepts $\mathcal{I}$– and $\mathcal{I}^*$–convergence coincide.

Theorem 1.11 (see, [20, 23]) *For any* $q \in (0, 1\rangle$ *the notions* $\mathcal{I}_c^{(q)}$– *and* $\mathcal{I}_c^{(q)*}$–*convergence are equivalent.*

Proof. It suffices to prove that for any $\mathcal{I}_c^{(q)}$, $q \in (0, 1\rangle$ and any sequence $(x_n)_{n=1}^{\infty}$ of real numbers such that $\mathcal{I}_c^{(q)} - \lim x_n = L$ for $q \in (0, 1\rangle$ there exists a set $M = \{m_1 < m_2 < \cdots < m_k < \cdots\} \subseteq \mathbb{N}$ such that $\mathbb{N} \backslash M \in \mathcal{I}_c^{(q)}$ and $\lim_{k \to \infty} x_{m_k} = L$.

For any positive integer k let $\varepsilon_k = \frac{1}{2^k}$ and $A_k = \left\{ n \in \mathbb{N} : |x_n - L| \geq \frac{1}{2^k} \right\}$. As $\mathcal{I}_c^{(q)} - \lim x_n = L$, we have $A_k \in \mathcal{I}_c^{(q)}$, i.e.

$$\sum_{a \in A_k} a^{-q} < \infty.$$

Therefore there exists an infinite sequence $n_1 < n_2 < \cdots < n_k < \cdots$ of integers such that for every $k = 1, 2, \dots$

$$\sum_{\substack{a>n_k \\ a\in A_k}} a^{-q} < \frac{1}{2^k}.$$

Let $H = \cup_{k=1}^{\infty}[(n_k, n_{k+1}\rangle \cap A_k]$. Then

$$\sum_{a\in H} a^{-q} \le \sum_{\substack{a>n_1 \\ a\in A_1}} a^{-q} + \sum_{\substack{a>n_2 \\ a\in A_2}} a^{-q} + \cdots + \sum_{\substack{a>n_k \\ a\in A_k}} a^{-q} + \cdots$$
$$< \frac{1}{2} + \frac{1}{2^2} + \cdots + \frac{1}{2^k} + \cdots < +\infty.$$

Thus $H \in \mathcal{I}_c^{(q)}$. Put $M = \mathbb{N}\setminus H = \{m_1 < m_2 < \cdots < m_k < \cdots\}$. Now it suffices to prove that $\lim_{k\to\infty} x_{m_k} = L$. Let $\varepsilon > 0$. Choose $k_0 \in \mathbb{N}$ such that $\frac{1}{2^{k_0}} < \varepsilon$. Let $m_k > n_{k_0}$. Then m_k belongs to some interval $(n_j, n_{j+1}\rangle$ where $j \ge k_0$ and does not belong to A_j ($j \ge k_0$). Hence m_k belongs to $\mathbb{N}\setminus A_j$, and then $|x_{m_k} - L| < \varepsilon$ for every $m_k > n_{k_0}$, thus $\lim_{k\to\infty} x_{m_k} = L$.

Corollary 1.12 *Ideals $\mathcal{I}_c^{(q)}$ for $q \in (0, 1\rangle$ have the property (AP).*

It is easy to prove the following lemma.

Lemma 1.13 (see [3]). *If $\mathcal{I}_1 \subseteq \mathcal{I}_2$ then the statement $\mathcal{I}_1 - \lim x_n = L$ implies $\mathcal{I}_2 - \lim x_n = L$.*

5. $\mathcal{I}$–convergence of arithmetical functions

We can obtain a good information about behaviour and properties of the well-known arithmetical functions by investigating $\mathcal{I}$–convergence of these functions or some sequences connected with these functions. Recall the concept of normal order.

Definition 1.14. The sequence $(x_n)_{n=1}^{\infty}$ has *the normal order* $(y_n)_{n=1}^{\infty}$ if for every $\varepsilon > 0$ and almost all (almost all in the sense of asymptotic density) values n we have $(1-\varepsilon)y_n < x_n < (1+\varepsilon)y_n$.

Schinzel and Šalát in [28] pointed out that one of equivalent definitions to have the normal order is as follows. The sequence $(x_n)_{n=1}^{\infty}$ has the normal order $(y_n)_{n=1}^{\infty}$ if and only if $\mathcal{I}_d - \lim \frac{x_n}{y_n} = 1$. The results concerning the normal order will be formulated using the concept of statistical convergence, which coincides with $\mathcal{I}_d$–convergence. For equivalent definitions of the normal order and more examples concerning this notion see [34, 38, 48].

In the papers [21, 27, 28] and in the monograph [38] there are studied various kinds of convergence of arithmetical functions which were mentioned at the beginning. The following equalities were proved in the paper [28] by using the concept of the normal order.

$$\mathcal{I}_d - \lim \frac{\omega(n)}{\log\log n} = \mathcal{I}_d - \lim \frac{\Omega(n)}{\log\log n} = 1$$

and

$$\mathcal{I}_d - \lim \frac{h(n)}{\log n} = \mathcal{I}_d - \lim \frac{H(n)}{\log n} = 0.$$

Similarly for the functions $f(n)$ and $f^*(n)$. In [27] it is proved the following equality:

$$\mathcal{I}_d - \lim \frac{\log \log f(n)}{\log \log n} = \mathcal{I}_d - \lim \frac{\log \log f^*(n)}{\log \log n} = 1 + \log 2.$$

Let us recall one more result from [26], let p be a prime number, there was proved that the sequence $\left(\log p \cdot \frac{a_p(n)}{\log n}\right)_{n=2}^{\infty}$ is $\mathcal{I}_d$–convergent to 0. Moreover the sequence $\left(\log p \cdot \frac{a_p(n)}{\log n}\right)_{n=2}^{\infty}$ is $\mathcal{I}_c^{(q)}$–convergent to 0 for $q=1$ and it is not $\mathcal{I}_c^{(q)}$–convergent for all $q \in (0,1)$, as it was shown in [21]. In [19] it was proved that this sequence is also
$\mathcal{I}_u$–convergent.

The following theorem shows that the assertions using the notion $\mathcal{I}_u$ instead of $\mathcal{I}_c^{(q)}$, $q \in (0,1\rangle$ need to use a different technique for their proofs. First of all we recall a new kind of convergence so called the uniformly strong ℓ–Cesàro convergence. This convergence is an analog of the notion of strong almost convergence (see [6]).

Definition 1.15. A sequence $(x_n)_{n=1}^{\infty}$ is said to be *uniformly strong ℓ–Cesàro convergent* $(0 < \ell < \infty)$ to a number L if $\lim_{N\to\infty} \frac{1}{N}\sum_{n=k+1}^{k+N} |x_i - L|^{\ell} = 0$ uniformly in k.

The following Theorem shows a connection between uniformly strong ℓ–Cesàro convergence and $\mathcal{I}_u$–convergence.

Theorem 1.16 (see [6]). *If $(x_n)_{n=1}^{\infty}$ is a bounded sequence, then $(x_n)_{n=1}^{\infty}$ is $\mathcal{I}_u$–convergent to L if and only if $(x_n)_{n=1}^{\infty}$ is uniformly strong ℓ–Cesàro convergent to L for some ℓ, $0 < \ell < \infty$.*

The sequence $\left(\log p \cdot \frac{a_p(n)}{\log n}\right)_{n=2}^{\infty}$ is $\mathcal{I}_u$–convergent to zero i.e. for arbitrary $\varepsilon > 0$ the set $A(\varepsilon) = \left\{n \in \mathbb{N} : \log p \cdot \frac{a_p(n)}{\log n} \geq \varepsilon > 0\right\}$ has uniform density equal to zero.

Theorem 1.17 (see [19]). *We have* $\mathcal{I}_u - \lim \log p \cdot \frac{a_p(n)}{\log n} = 0$.

Proof. The sequence $\left(\log p \cdot \frac{a_p(n)}{\log n}\right)_{n=2}^{\infty}$ is bounded. Using Theorem 1.16, it is sufficient to show that the sequence $\left(\log p \cdot \frac{a_p(n)}{\log n}\right)_{n=2}^{\infty}$ is uniformly strong ℓ–Cesàro convergent to 0 for $\ell = 1$. For the reason that all members of $\left(\log p \cdot \frac{a_p(n)}{\log n}\right)_{n=2}^{\infty}$ are positive, we shall prove that $\lim_{N\to\infty} \frac{1}{N}\sum_{n=k+1}^{k+N} \frac{a_p(n)}{\log n} = 0$, uniformly in k. $a_p(n) = \alpha$ if $p^{\alpha} \| n$. Let $\alpha_0 = \left[\frac{\log N}{\log p}\right]$. This immediately implies that $p^{\alpha_0} \leq N < p^{\alpha_0+1}$. Then for all $n \in (k, k+N]$ we have $a_p(n) = \alpha < \alpha_0$ with the possible exception of one $n_1 \in (k, k+N]$ for which we could have $a_p(n_1) = \alpha_1 > \alpha_0$. Assume that there exist two such numbers $n_1, n_2 \in (k, k+N]$ for which $a_p(n_1) = \alpha_1 > \alpha_0$ and $a_p(n_2) = \alpha_2 > \alpha_0$, then $n_1 = m_1 p^{\alpha_1}$, $n_2 = m_2 p^{\alpha_2}$ hence $p^{\alpha_0+1} | n_1 - n_2$. We have p^{α_0+1} $< |n_1 - n_2| \leq N$, what is a contradiction with $p^{\alpha_0+1} > N$. When we omit such an n_1 from the sum, the error is less than $\frac{1}{N}\frac{a_p(n_1)}{\log n_1} \leq \frac{1}{N}\frac{\alpha_1}{\alpha_1} \log p$. Using the Hölder's inequality we get

$$\frac{1}{N}\sum_{n=k+1}^{k+N} \frac{a_p(n)}{\log n} \leq \sqrt{\frac{1}{N}\sum_{n=k+1}^{k+N} (a_p(n))^2} \sqrt{\frac{1}{N}\sum_{n=k+1}^{k+N} \frac{1}{(\log n)^2}}. \quad (3)$$

We are going to estimate the first factor of Eq. (3)

$$\frac{1}{N}\sum_{n=k+1}^{k+N}\left(a_p(n)\right)^2 \cong \frac{1}{N}\sum_{\alpha=0}^{\alpha_0}\alpha^2 \sum_{\substack{n=k+1\\ a_p(n)=\alpha}}^{k+N} 1 =$$
$$\frac{1}{N}\sum_{\alpha=0}^{\alpha_0}\alpha^2\left(\left[\frac{k+N}{p^\alpha}\right]-\left[\frac{k}{p^\alpha}\right]-\left(\left[\frac{k+N}{p^{\alpha+1}}\right]-\left[\frac{k}{p^{\alpha+1}}\right]\right)\right)=$$
$$\frac{1}{N}\sum_{\alpha=0}^{\alpha_0}\alpha^2\left(\frac{N}{p^\alpha}-\frac{N}{p^{\alpha+1}}+O(1)\right)=\frac{1}{N}N\left(1-\frac{1}{p}\right)\sum_{\alpha=0}^{\alpha_0}\alpha^2\frac{1}{p^\alpha}+\frac{O(1)}{N}\sum_{\alpha=0}^{\alpha_0}\alpha^2.$$

Formula $\sum_{\alpha=0}^{\alpha_0}\alpha^2 = P(\alpha_0)$, where $P(x)=\frac{x(x+1)(2x+1)}{6}$ and simple estimations give $\sum_{\alpha=0}^{\alpha_0}\alpha^2\frac{1}{p^\alpha}\leq\sum_{\alpha=0}^{\infty}\frac{\alpha^2}{p^\alpha}<\infty$.

So we get

$$\frac{1}{N}\sum_{n=k+1}^{k+N}\left(a_p(n)\right)^2 = O(1). \tag{4}$$

Estimate the second factor Eq. (3)

$$\frac{1}{N}\sum_{n=k+1}^{k+N}\frac{1}{(\log n)^2}\leq\frac{1}{N}\sum_{n=2}^{N+1}\frac{1}{(\log n)^2}\to 0,\quad \text{since}\quad \frac{1}{(\log n)^2}\to 0. \tag{5}$$

Let $N\to\infty$, from Eqs. (4) and (5) we obtain

$$\frac{1}{N}\sum_{n=k+1}^{k+N}\frac{a_p(n)}{\log n}\leq c\sqrt{\frac{1}{N}\sum_{n=2}^{N+1}\frac{1}{(\log n)^2}}\to 0,$$

uniformly in k.

Remark. It is known that $\mathcal{I}_u \subsetneq \mathcal{I}_d$ (see e.g. [5, 6]) but the ideals $\mathcal{I}_c$ and $\mathcal{I}_u$ are not disjoint, and moreover $\mathcal{I}_u \not\subseteq \mathcal{I}_c$ and $\mathcal{I}_c \not\subseteq \mathcal{I}_u$. For example the set of all prime numbers belongs to $\mathcal{I}_u$ but not belongs to $\mathcal{I}_c$. On the other hand there exists the set $B=\cup_{k=1}^{\infty}B_k$, where $B_k=\{k^3+1,k^3+2,\dots,k^3+k\}$ which not belongs to $\mathcal{I}_u$ but it belongs to $\mathcal{I}_c$.

Under the fact that $\mathcal{I}_c^{(q)} \subsetneq \mathcal{I}_d$ for all $q\in(0,1\rangle$ and Lemma 1.13 it is useful to investigate $\mathcal{I}_c^{(q)}$–convergence of special sequences described in the introduction. Under the Lemma 1.13 it is clear that if there exists the $\mathcal{I}_c^{(q)}$–limit of some sequence for any $q\in(0,1\rangle$, then it is equal to the $\mathcal{I}_d$–limit of the same sequence. There are no other options.

Consider the sequences $\left(\frac{h(n)}{\log n}\right)_{n=2}^{\infty}$ and $\left(\frac{H(n)}{\log n}\right)_{n=2}^{\infty}$. In [28] it was proved that these sequences are dense on $\left(0,\frac{1}{\log 2}\right)$ and moreover they both are statistically convergent to zero. The same result we have for $\mathcal{I}_c^{(q)}$–convergence, but only for the sequence $\left(\frac{h(n)}{\log n}\right)_{n=2}^{\infty}$ for all $q\in(0,1\rangle$.

Theorem 1.18 (see [20]). *We have*

$$\mathcal{I}_c^{(q)}-\lim\frac{h(n)}{\log n}=0,\quad \textit{for all}\quad q\in(0,1\rangle.$$

Proof. Let $k \in \mathbb{N}$ and $k \geq 2$. It is easy to see that the following equality holds

$$1 + \sum_{\substack{n \in \mathbb{N} \\ h(n) \geq k}} n^{-q} = \prod_{p \in \mathbb{P}} \left(1 + \frac{1}{p^{kq}} + \frac{1}{p^{(k+1)q}} + \cdots\right), \tag{6}$$

where $\mathbb{P}$ denotes the set of all primes.
The right-hand side of the equality Eq. (6) equals

$$\prod_{p \in \mathbb{P}} \left(1 + \frac{1}{p^{kq}} \cdot \frac{1}{1 - \frac{1}{p^q}}\right) = \prod_{p \in \mathbb{P}} \left(1 + \frac{1}{p^{(k-1)q} \cdot (p^q - 1)}\right).$$

Then for $q > \frac{1}{k}$, the product on the right-hand side of the previous equality converges. Thus, the series on the left-hand side of Eq. (6) converges.

Let $\varepsilon > 0$. Put $A(\varepsilon) = \left\{n \in \mathbb{N} : \frac{h(n)}{\log n} \geq \varepsilon > 0\right\}$. There exists an $n_0^{(k)} \in \mathbb{N}$ for all $k \geq 2$ such that for all $n > n_0^{(k)}$ and $n \in A(\varepsilon)$ we have $h(n) \geq \varepsilon \cdot \log n > k$ (it is sufficient to put $n_0^{(k)} = \left[e^{\frac{k}{\varepsilon}}\right]$).

From this $A(\varepsilon) \cap \left\{n_0^{(k)} + 1, n_0^{(k)} + 2, \ldots\right\} \subseteq \{n \in \mathbb{N} : h(n) \geq k\}$ for all $k \geq 2, k \in \mathbb{N}$.

Therefore $\sum_{n \in A(\varepsilon)} n^{-q} < +\infty$ for all $k \geq 2$ and $\mathcal{I}_c^{(q)} - \lim \frac{h(n)}{\log n} = 0$ since the series Eq. (6) converges for all $q > \frac{1}{k}$. If $k \to \infty$ for sufficient large then $\mathcal{I}_c^{(q)} - \lim \frac{h(n)}{\log n} = 0$ for all $q \in (0, 1\rangle$.

Corollary 1.19. We have

$$\mathcal{I}_c^{(q)*} - \lim \frac{h(n)}{\log n} = 0 \ \ for \ \ all \ \ q \in (0, 1\rangle.$$

For the sequence $\left(\frac{H(n)}{\log n}\right)_{n=2}^{\infty}$ we get the result of different character.

Theorem 1.20 (see [20]). *The sequence* $\left(\frac{H(n)}{\log n}\right)_{n=2}^{\infty}$ *is not* $\mathcal{I}_c^{(q)}$*–convergent for every* $q \in (0, 1)$.

Proof. In the paper [21] is proved, that the sequence $\left(\log p \cdot \frac{a_p(n)}{\log n}\right)_{n=2}^{\infty}$ is not $\mathcal{I}_c^{(q)}$–convergent for any $q \in (0, 1)$. The sequence $\left(\frac{a_p(n)}{\log n}\right)_{n=2}^{\infty}$ is also not $\mathcal{I}_c^{(q)}$–convergent to zero. The inequality $H(n) \geq a_p(n)$ holds for all $n = 1, 2, \ldots$ and for any prime number p. Then we have $\frac{H(n)}{\log n} \geq \frac{a_p(n)}{\log n}$ for all $n = 2, 3, \ldots$. This implies that the sequence $\left(\frac{H(n)}{\log n}\right)_{n=2}^{\infty}$ is also not $\mathcal{I}_c^{(q)}$–convergent to zero for every $q \in (0, 1)$.

Theorem 1.21 (see [20]). *For* $q = 1$, *we obtain*

$$\mathcal{I}_c - \lim \frac{H(n)}{\log n} = 0.$$

Proof. We will show that

$$A(\varepsilon) = \left\{n \in \mathbb{N} : \frac{H(n)}{\log n} \geq \varepsilon\right\} \in \mathcal{I}_c$$

for any $\varepsilon > 0$.

Every non-negative integer n can be represented as $n = ab^2$, where a is a square-free number. Hence $H(a) = 1$ and

$$H(n) \in \{H(b^2), H(b^2) + 1\}.$$

If $n \in A(\varepsilon)$ then from $H(n) \geq \varepsilon \cdot \log n$ we have

$$\log (ab^2) \leq \frac{H(b^2) + 1}{\varepsilon} \text{ and so } \log a \leq \frac{H(b^2) + 1}{\varepsilon}.$$

Therefore

$$A(\varepsilon) \subseteq B = \left\{ n \in \mathbb{N} : n = ab^2, \ \ \log a \leq \frac{H(b^2) + 1}{\varepsilon}, \ \ b \in \mathbb{N} \right\}.$$

It is enough to prove that $\sum_{n \in B} n^{-1} < +\infty$. We have

$$\sum_{n \in B} \frac{1}{n} = \sum_{b=1}^{\infty} \frac{1}{b^2} \sum_{\log a \leq \frac{H(b^2)+1}{\varepsilon}} \frac{1}{a}.$$

We use the inequality $S_k = \sum_{j=1}^{k} \frac{1}{j} \leq 1 + \log k$ for the harmonic series. Then we have the following inequality

$$\sum_{n \in B} \frac{1}{n} \leq \sum_{b=1}^{\infty} \frac{1}{b^2} \left(\frac{H(b^2) + 1}{\varepsilon} + 1 \right). \tag{7}$$

Because the $\sum \frac{1}{b^2} = \frac{\pi^2}{6} < +\infty$, it is enough to prove that the

$$\sum_{b=1}^{\infty} \frac{H(b^2)}{b^2} < +\infty. \tag{8}$$

For any $n \in \mathbb{N}$ we have $n = p_1^{a_1} \cdots p_k^{a_k} \geq 2^{H(n)}$ and from this $H(n) \leq \frac{\log n}{\log 2}$. Therefore

$$\sum_{b=1}^{\infty} \frac{H(b^2)}{b^2} \leq \frac{2}{\log 2} \sum_{b=1}^{\infty} \frac{\log b}{b^2} < +\infty.$$

We have shown that the sum in Eq. (8) is finite and therefore the sum in Eq. (7) is also finite.

Moreover $B \in \mathcal{I}_c$ and because $A(\varepsilon) \subseteq B$ we have $A(\varepsilon) \in \mathcal{I}_c$.

The situation for sequences $\left(\frac{\omega(n)}{\log \log n} \right)_{n=2}^{\infty}$, $\left(\frac{\Omega(n)}{\log \log n} \right)_{n=2}^{\infty}$ is following.

Theorem 1.22 (see [20]). *The sequences* $\left(\frac{\omega(n)}{\log \log n} \right)_{n=2}^{\infty}$ *and* $\left(\frac{\Omega(n)}{\log \log n} \right)_{n=2}^{\infty}$ *are not* $\mathcal{I}_c^{(q)}$*–convergent for all* $q \in (0, 1\rangle$.

Proof. We prove this assertion only for $\left(\frac{\omega(n)}{\log \log n} \right)_{n=2}^{\infty}$. The proof for the sequence $\left(\frac{\Omega(n)}{\log \log n} \right)_{n=2}^{\infty}$ is analogous. Let $q = 1$. On the basis of the Theorem 2.2 of [28] and Lemma 1.13 we can assume that $\mathcal{I}_c - \lim \frac{\omega(n)}{\log \log n} = 1$. Take $\varepsilon \in (0, \frac{1}{2})$ and consider the set

$$A(\varepsilon) = \left\{ n \in \mathbb{N} : \left| \frac{\omega(n)}{\log \log n} - 1 \right| \geq \varepsilon \right\}.$$

Put $n = p$, where p is a prime number, then $\omega(p) = 1$ and $\left| \frac{1}{\log \log p} - 1 \right| \geq \varepsilon$ holds for all prime numbers $p > p_0$. Therefore the set A_ε contains all prime numbers greater than p_0. For these p we have: $\sum_{p > p_0} \frac{1}{p} = +\infty$ and so $A(\varepsilon) \notin \mathcal{I}_c$. From this $\mathcal{I}_c - \lim \frac{\omega(n)}{\log \log n} \neq 1$. Under the inclusion $\mathcal{I}_c^{(q)} \subsetneq \mathcal{I}_c^{(1)} \equiv \mathcal{I}_c$ and according to Lemma 1.13 we have $\mathcal{I}_c^{(q)} - \lim \frac{\omega(n)}{\log \log n} \neq 1$ for $q \in (0, 1\rangle$. This complete the proof.

Further possibility where the results can be strengthened by the way that the statistical convergence in them is replaced by $\mathcal{I}_c^{(q)}$–convergence is the concept of the famous *Pascal's triangle*. The n-th row of the Pascal's triangle consists of the numbers $\binom{n}{0}, \binom{n}{1}, \ldots, \binom{n}{n-1}, \binom{n}{n}$. Their sum equals to $2^n = (1+1)^n = \sum_{k=0}^{n} \binom{n}{k}$. Let $\Gamma(t)$ denote the number of times the positive integer t, $t > 2$ occurs in the Pascal's triangle. That is, $\Gamma(t)$ is the number of binomial coefficients $\binom{n}{k}$ satisfying $\binom{n}{k} = t$. From this point of view Γ is the function which maps the set $\mathbb{N}$ in the set $\mathbb{N} \cup \{\aleph_0\}$ $(\Gamma(1) = \aleph_0)$. Let us observe that for every $t \in \mathbb{N}$, $\Gamma(t) \geq 1$.

In [32] it is proved that the average and normal order of the function Γ is 2. Since the normal order is 2, we have

$$\mathcal{I}_d - \lim \Gamma(t) = 2$$

(see [28]). We are going to show two results which strengthen the result of [32] and their proofs are outlined in [24].

Theorem 1.23 (see [24]). $\mathcal{I}_c - \lim \Gamma(t) = 2$.

Proof. The values of the function Γ are positive integers for $t \neq 1$. Thus for $\varepsilon > 0$ the set $A_\varepsilon = \{t \in \mathbb{N} : |\Gamma(t) - 2| \geq \varepsilon\}$ is a subset of the set $H = \{1\} \cup \{2\} \cup M$, where $M = \{t \in \mathbb{N} : \Gamma(t) > 2\}$. Note that $\Gamma(2) = 1$. Therefore is suffices to show that $\sum_{n \in H} \frac{1}{n} < +\infty$. Evidently this is equivalent with

$$\sum_{n \in M} \frac{1}{n} < +\infty. \tag{9}$$

We shall prove Eq. (9). Firstly, we write the left-hand site of Eq. (9) in the form

$$\sum_{n \in M} \frac{1}{n} = \frac{M(1)}{1} + \frac{M(2) - M(1)}{2} + \cdots + \frac{M(k) - M(k-1)}{k} + \cdots$$

$$= \frac{M(1)}{1 \cdot 2} + \frac{M(2)}{2 \cdot 3} + \cdots + \frac{M(k)}{k \cdot (k+1)} + \cdots.$$

In [32] it is shown that $M(x) = O(\sqrt{x})$. Therefore there exists such $c_1 > 0$ that for every $k \in \mathbb{N}$, $M(k) \leq c_1 \sqrt{k}$ holds. But then

$$\frac{M(k)}{k \cdot (k+1)} \leq \frac{c_1}{k^{\frac{3}{2}}} \quad (k = 1, 2, \ldots).$$

According these inequalities by comparison test of the convergence of the series in Eq. (9) follows.

Now we shall use the concept of $\mathcal{I}_c^{(q)}$–convergence.

Theorem 1.24 (see [24]). *For every* $q > \frac{1}{2}$, $\mathcal{I}_c^{(q)} - \lim \Gamma(t) = 2$ *holds and* $\mathcal{I}_c^{(q)} - \lim \Gamma(t) = 2$ *does not hold for any* q, $0 < q \le \frac{1}{2}$.

Proof. Let $0 < q < 1$ and let M have the same meaning as in the proof of Theorem 1.23. Let us examine the series $\sum_{n \in M} \frac{1}{n^q}$. We write it in the form

$$\begin{aligned} \sum_{n \in M} \frac{1}{n^q} &= \frac{M(1)}{1^q} + \frac{M(2) - M(1)}{2^q} + \cdots + \frac{M(k) - M(k-1)}{k^q} + \cdots \\ &= M(1)\left(\frac{1}{1^q} - \frac{1}{2^q}\right) + \cdots + M(k)\left(\frac{1}{k^q} - \frac{1}{(k+1)^q}\right) + \cdots \\ &= \sum_{k=1}^{\infty} M(k)\left(\frac{1}{k^q} - \frac{1}{(k+1)^q}\right). \end{aligned} \tag{10}$$

In virtue of Lagrange's mean value theorem we have

$$(k+1)^q - k^q = q z_k^{q-1}, \quad k < z_k < k+1 \quad (k = 1, 2, \dots).$$

Therefore the series Eq. (10) can be written in the form

$$\sum_{k \in M} \frac{1}{k^q} = \sum_{k=1}^{\infty} \frac{M(k) q z_k^{q-1}}{k^q (k+1)^q} = \sum_{k=1}^{\infty} \frac{q M(k)}{k^q (k+1)^q z_k^{1-q}}. \tag{11}$$

But $z_k^{1-q} > k^{1-q}$, $(k+1)^q > k^q$ and so

$$\sum_{n \in M} \frac{1}{n^q} \le q \sum_{k=1}^{\infty} \frac{M(k)}{k^{1+q}}.$$

We have already seen, that $M(k) \le c_1 \sqrt{k}$, $(k = 1, 2, \dots)$ (in the proof of Theorem 1.23). Consider that every binomial coefficient $t = \binom{n}{2}$, $n \ge 4$ occurs in Pascal's triangle at least four times $\left(\text{as } \binom{n}{2}, \binom{n}{n-2}, \binom{t}{1}, \binom{t}{t-1}\right)$. Therefore every number of this form belongs to M. Consequently for $x > 4$, $x \in \mathbb{N}$ the number $M(x)$ is greater then or equal to the number $V(x)$ of all numbers of the form $\binom{n}{2}$, $n \ge 4$ not exceeding x. But $V(x) \ge s - 3$, where s is the integer satisfying

$$\binom{s}{2} \le x < \binom{s+1}{2}.$$

From this we get

$$x < \frac{s(s+1)}{2}, \quad s > \sqrt{2x} - 1.$$

So we obtain

$$M(x) \geq V(x) \geq s - 3 > \sqrt{2x} - 4, \qquad \frac{M(x)}{\sqrt{x}} > \sqrt{2} - \frac{4}{x} \geq \underbrace{\sqrt{2} - 1}_{c_2} > 0.$$

Now it is clear that if $x \geq \binom{4}{2}$ then

$$c_2\sqrt{x} \leq M(x) \leq c_1\sqrt{x}, \quad c_2 = \sqrt{2} - 1 > 0. \tag{12}$$

Therefore by Eq. (11) we get

$$\sum_{n \in M} \frac{1}{n^q} = q \sum_{k=1}^{\infty} \frac{M(k)}{k^q (k+1)^q z_k^{1-q}}.$$

But $z_k < k+1$, hence $z_k^{1-q} < (k+1)^{1-q}$ and so

$$\sum_{n \in M} \frac{1}{n^q} \geq q \sum_{k=1}^{\infty} \frac{M(k)}{(k+1)^{1+q}}.$$

From this owing to Eq. (12) we obtain

$$\sum_{n \in M} \frac{1}{n^q} \geq q \sum_{k=1}^{\infty} \frac{c_2\sqrt{k}}{(k+1)^{1+q}} \geq q \frac{c_2}{2} \sum_{k=1}^{\infty} \frac{1}{(k+1)^{\frac{1}{2}+q}} = +\infty \quad \text{if} \quad 0 < q \leq \frac{1}{2}.$$

Thus $\sum_{n \in M} \frac{1}{n^q} = +\infty$, and so $\sum_{n \in A_\varepsilon} \frac{1}{n^q} = +\infty$ for every $\varepsilon > 0$.

Similar results we can prove for functions $f(n)$ and $f^*(n)$.

Theorem 1.25 (see [20, 27]). *The sequence* $\left(\frac{\log \log f(n)}{\log \log n}\right)_{n=2}^{\infty}$ *is not* $\mathcal{I}_c^{(q)}$*–convergent for all* $q \in (0, 1\rangle$.

Proof. According to Theorem 2.1 of [27] suppose that the

$$\mathcal{I}_c^{(q)} - \lim \frac{\log \log f(n)}{\log \log n} = 1 + \log 2,$$

where $q \in (0, 1\rangle$. Let $\varepsilon \in (0, \log 2)$ and define the set

$$A(\varepsilon) = \left\{ n \in \mathbb{N} : \left| \frac{\log \log f(n)}{\log \log n} - (1 + \log 2) \right| \geq \varepsilon \right\}.$$

Put $n = p$, where p is a prime number, then $f(p) = p$ and $\frac{\log \log p}{\log \log p} = 1$. Therefore the set $A(\varepsilon)$ contains all prime numbers. Next we have:

$$\sum_{n \in A(\varepsilon)} n^{-q} \geq \sum_{j=1}^{\infty} p_j^{-q} \geq \sum_{j=1}^{\infty} p_j^{-1} = +\infty, \quad q \in (0, 1\rangle.$$

Hence $A(\varepsilon) \notin \mathcal{I}_c^{(q)}$ and $\mathcal{I}_c^{(q)} - \lim \frac{\log \log f(n)}{\log \log n} \neq 1 + \log 2$ for all $q \in (0, 1\rangle$.

Theorem 1.26 (see [20, 27]). *The sequence* $\left(\frac{\log \log f^*(n)}{\log \log n}\right)_{n=2}^{\infty}$ *is not* $\mathcal{I}_c^{(q)}$*–convergent for all* $q \in (0, 1\rangle$.

Proof. According to Theorem 2.2 of [27] again suppose that the

$$\mathcal{I}_c^{(q)} - \lim \frac{\log \log f^*(n)}{\log \log n} = 1 + \log 2,$$

where $q \in (0, 1\rangle$. The proof is going similar as in the previous Theorem. Put $n = p_i p_j$, $i \neq j$, where p_i, p_j are distinct prime numbers. Then $f^*(n) = f^*\left(p_i p_j\right) = \frac{f(p_i p_j)}{p_i p_j} = \frac{p_i p_j (p_i p_j)}{p_i p_j} = p_i p_j, i \neq j$. Hence $\frac{\log \log f^*(p_i p_j)}{\log \log p_i p_j} = 1$. Let $\varepsilon \in (0, \log 2)$ and define the set

$$A(\varepsilon) = \left\{ n \in \mathbb{N} : \left| \frac{\log \log f^*(n)}{\log \log n} - (1 + \log 2) \right| \geq \varepsilon \right\}.$$

This set contains all numbers of the type $p_i p_j$, $i \neq j$. For $q \in (0, 1\rangle$ we have:

$$\sum_{n \in A(\varepsilon)} n^{-q} \geq \sum_{\substack{j=1 \\ p_j \neq 2}}^{\infty} \frac{1}{2p_j}, \quad (p_i = 2).$$

Since the series $\sum_{j=1}^{\infty} \frac{1}{2p_j}$ diverges, we have $A(\varepsilon) \notin \mathcal{I}_c^{(q)}$ for all $q \in (0, 1\rangle$. Therefore $\mathcal{I}_c^{(q)} - \lim \frac{\log \log f^*(n)}{\log \log n} \neq 1 + \log 2$ and the proof is complete.

There exists a relationship between functions $f(n)$ and $d(n)$ (where $d(n)$ is the number of divisors of n). The following equality holds: $\log f(n) = \frac{d(n)}{2} \cdot \log n$, $(n > e)$ (see [34]). From this we have

$$\log \log f(n) = \log \frac{1}{2} + \log d(n) + \log \log n, \quad n > e^e.$$

Therefore

$$\frac{\log \log f(n)}{\log \log n} = 1 + \frac{\log d(n)}{\log \log n} + \frac{\log \frac{1}{2}}{\log \log n}, \quad n > e^e.$$

From Theorem 1.25 we have the following statement.

Corollary 1.27. *The sequence* $\left(\frac{\log d(n)}{\log \log n}\right)_{n=2}^{\infty}$ *is not* $\mathcal{I}_c^{(q)}$*–convergent for all* $q \in (0, 1\rangle$.

The following results concerning the functions $\gamma(n)$ and $\tau(n)$.

In [33, Theorem 3, 5] there are proofs of the following results:

$$\sum_{n=2}^{\infty} \frac{\gamma(n) - 1}{n} = 1, \qquad \sum_{n=2}^{\infty} \frac{\tau(n) - 1}{n} = 1 + \frac{\pi^2}{6}.$$

In connection with these results we have investigated the convergence of series for any $\alpha \in (0, 1)$,

$$\sum_{n=2}^{\infty} \frac{\gamma(n) - 1}{n^{\alpha}}, \qquad \sum_{n=2}^{\infty} \frac{\tau(n) - 1}{n^{\alpha}},$$

that we need for $\mathcal{I}_c^{(q)}$–convergence of functions $\gamma(n)$ and $\tau(n)$. The following results are outlined in [21].

Theorem 1.28. *The series*

$$\sum_{n=2}^{\infty} \frac{\gamma(n)-1}{n^{\alpha}}$$

diverges for $0 < \alpha \leq \frac{1}{2}$ and converges for $\alpha > \frac{1}{2}$.
Proof.

a. Let $0 < \alpha \leq \frac{1}{2}$. Put $K = \{k^2 : k \in \mathbb{N}, k > 1\}$. A simple estimation gives

$$\sum_{n=2}^{\infty} \frac{\gamma(n)-1}{n^{\alpha}} \geq \sum_{n \in K} \frac{\gamma(n)-1}{n^{\alpha}}.$$

Clearly $\gamma(n) \geq 2$ for $n \in K$. Therefore

$$\sum_{n=2}^{\infty} \frac{\gamma(n)-1}{n^{\alpha}} \geq \sum_{n \in K} \frac{1}{n^{\alpha}} = \sum_{k=2}^{\infty} \frac{1}{k^{2\alpha}} \geq \sum_{k=2}^{\infty} \frac{1}{k} = +\infty. \tag{13}$$

b. Let $\alpha > \frac{1}{2}$. We will use the formula

$$\sum_{n=2}^{\infty} \frac{\gamma(n)-1}{n^{\alpha}} = \sum_{k=2}^{\infty}\sum_{s=2}^{\infty} \frac{1}{k^{\alpha s}} = \sum_{k=2}^{\infty} \frac{1}{k^{\alpha}(k^{\alpha}-1)}. \tag{14}$$

For a sufficiently large number k $(k > k_0)$ we have $\frac{k^{\alpha}}{k^{\alpha}-1} < 2$. We can estimate the series on the right-hand side of Eq. (14) with

$$\sum_{k=2}^{\infty} \frac{1}{k^{\alpha}(k^{\alpha}-1)} < \sum_{k=2}^{k_0} \frac{1}{k^{\alpha}(k^{\alpha}-1)} + 2\sum_{k > k_0} \frac{1}{k^{2\alpha}}.$$

Since $2\alpha > 1$ we get

$$\sum_{n=2}^{\infty} \frac{\gamma(n)-1}{n^{\alpha}} < +\infty.$$

Corollary 1.29. *The sequence* $\gamma(n)$ *is*

i. $\mathcal{I}_c$–convergent to 1,

ii. $\mathcal{I}_c^{(q)}$–divergent for $q \in \left(0, \frac{1}{2}\right\rangle$ and $\mathcal{I}_c^{(q)}$–convergent to 1 for $q \in \left(\frac{1}{2}, 1\right)$.

Proof.

i. Let $\varepsilon > 0$. The set of numbers $\{n \in \mathbb{N}, n > 1 : |\gamma(n) - 1| \geq \varepsilon\}$ is a subset of $H = \{n = t^s : n \in \mathbb{N}, t > 1, s > 1\}$ and $\sum_{a \in H} \frac{1}{a} < +\infty$. From the definition of $\mathcal{I}_c$–convergence Cor. 1.29 i. (Cor. is the abbreviation for Corollary) follows.

ii. Let $\varepsilon > 0$ and denote $A_{\varepsilon} = \{n \in \mathbb{N} : |\gamma(n) - 1| \geq \varepsilon\}$. When $0 < q \leq \frac{1}{2}$ then for the numbers $n \in K$, $K = \{k^2 : k \in N, k > 1\}$ considering Eq. (13) for $q = \alpha$ holds

$$\sum_{n \in A_{\varepsilon}} \frac{1}{n^{\alpha}} \geq \sum_{n \in K} \frac{1}{n^{\alpha}} \geq +\infty.$$

Therefore $\gamma(n)$ is $\mathcal{I}_c^{(q)}$–divergent. If $\frac{1}{2} < q < 1$, $q = \alpha$ then $A_{\varepsilon} \subset H$ and

$$\sum_{n=2}^{\infty}\frac{1}{n^{\alpha}} \leq \sum_{k=2}^{\infty}\sum_{s=2}^{\infty}\frac{1}{k^{\alpha s}}.$$

The convergence of the series on the right-hand side we proved previously in Theorem 1.28. Therefore $\gamma(n)$ is $\mathcal{I}_c^{(q)}$–convergent to 1 if $q \in \left(\frac{1}{2}, 1\right)$.

Remark. We have $\lim \text{ stat } \gamma(n) = 1$.

Theorem 1.30 The series

$$\sum_{n=2}^{\infty}\frac{\tau(n)-1}{n^{\alpha}}$$

diverges for $0 < \alpha \leq \frac{1}{2}$ and converges for $\alpha > \frac{1}{2}$.

Proof. Let $0 < \alpha < 1$. We write the given series in the form

$$\sum_{n=2}^{\infty}\frac{\tau(n)-1}{n^{\alpha}} = \sum_{k=2}^{\infty}\sum_{s=2}^{\infty}\frac{s}{k^{\alpha s}}, \tag{15}$$

We shall try to use a similar method to Mycielski's proof of the convergence of $\sum_{n=2}^{\infty}\frac{\tau(n)-1}{n^{\alpha}}$ to explain the equality Eq. (15). Since $\frac{s}{k^{\alpha s}} = -\frac{k}{\alpha}\frac{d}{dt}\left(\frac{1}{t^{\alpha s}}\right)_{t=k}$ and $\sum_{s=2}^{\infty}\frac{1}{t^{\alpha s}} = \frac{1}{t^{\alpha}(t^{\alpha}-1)}$ the right-hand side of Eq. (15) is equal to

$$\sum_{s=2}^{\infty}\frac{2k^{\alpha}-1}{k^{\alpha}(k^{\alpha}-1)^2} = \sum_{s=2}^{\infty} a_k$$

For the k-th term of $\sum a_k$ we have

$$a_k = \frac{2-\frac{1}{k^{\alpha}}}{\left(1-\frac{1}{k^{\alpha}}\right)^2}\cdot\frac{1}{k^{2\alpha}}.$$

Denote by $b_k = \frac{1}{k^{2\alpha}}$ and consider that $\lim_{k\to\infty}\frac{a_k}{b_k} = 2$. Hence the series $\sum_{s=2}^{\infty}a_k$ converges (diverges) if and only if the series $\sum_{s=2}^{\infty}b_k$ converges (diverges). Since $\sum b_k$ is convergent (divergent) for any $\alpha > \frac{1}{2}$ $\left(0 < \alpha \leq \frac{1}{2}\right)$ so does the series $\sum a_k$ and therefore the series $\sum\frac{\tau(n)-1}{n^{\alpha}}$.

Corollary 1.31. *The sequence $\tau(n)$ is*

i. *$\mathcal{I}_c$–convergent to 1,*

ii. *$\mathcal{I}_c^{(q)}$–divergent for $q \in \left(0, \frac{1}{2}\right\rangle$ and $\mathcal{I}_c^{(q)}$–convergent to 1 for $q \in \left(\frac{1}{2}, 1\right)$.*

Proof. Similar to the proof of Corollary 1.29.

Remark. We have $\lim \text{ stat } \tau(n) = 1$.

6. Conclusions

It turns out that the study of $\mathcal{I}$–convergence of arithmetical functions or some sequences related to these arithmetical functions for different kinds of ideals $\mathcal{I}$

(see [18]) gives a deeper insight into the behaviour and properties of these arithmetical functions.

On the other hand Algebraic number theory has many deep applications in cryptology. Many basic algorithms, which are widely used, have its security due to ANT. The theory of arithmetic functions has many connections to the classical ciphers, and to the general theory as well.

Acknowledgements

This part was partially supported by The Slovak Research and Development Agency under the grant VEGA No. 2/0109/18.

Author details

Vladimír Baláž*† and Tomáš Visnyai†
Institute of Information Engineering, Automation, and Mathematics, Faculty of Chemical and Food Technology STU in Bratislava, Bratislava, Slovakia

*Address all correspondence to: vladimir.balaz@stuba.sk

† These authors contributed equally.

References

[1] Fast H. Sur la convergence statistique. Colloquia Mathematica. 1951;**2**(3–4):241-244

[2] Schoenberg IJ. The Integrability of certain functions and related Summability methods. American Mathematical Monthly. 1959;**66**(5): 361-375

[3] Kostyrko P, Wilczyński W, Šalát T. $\mathcal{I}$–convergence. Real Analysis Exchange. Bratislava; 2000;**26**(2): 669-686

[4] Bourbaki N. Éléments de mathématique: Topologie générale. In: Livre III, (Russian translation) Obščaja topologija. Moskow, Nauka: Osnovnye struktury; 1968

[5] Baláž V, Strauch O, Šalát T. Remarks on several types of convergence of bounded sequences. Acta Mathematica Universitatis Ostraviensis. 2006;**14**(1): 3-12

[6] Baláž V, Šalát T. Uniform density u and corresponding $\mathcal{I}_u$–convergence. Mathematical Communications. 2006; **11**(1):1-7

[7] Connor J. The statistical and strong p-Cesaro convergence of sequences. Analysis. 1988;**8**(1–2):47-64

[8] Fridy JA. On statistical convergence. Analysis. 1985;**5**(4):301-314

[9] Kostyrko P, Mačaj M, Šalát T, Sleziak M. $\mathcal{I}$–convergence and extremal $\mathcal{I}$–limit points. Mathematica Slovaca. 2005;**55**(4):443-464

[10] Fridy J, Miller H. A matrix characterization of statistical convergence. Analysis. 1991;**11**(1):59-66

[11] Filipów R, Tryba J. Representation of ideal convergence as a union and intersection of matrix summability methods. Journal of Mathematical Analysis and Applications. 2020;**484**(2). [Accessed: 15 April 2020]

[12] Giuliano R, Grekos G. On the upper and lower exponential density functions. Mathematica Slovaca. 2017; **67**(5):1105-1128

[13] Paštéka M, Šalát T, Visnyai T. Remarks on Buck's measure density and a generalization of asymptotic density. Tatra Mountains Mathematical Publications. 2005;**31**(2):87-101

[14] Kostyrko P, Mačaj M, Šalát T. Statistical convergence and $\mathcal{I}$–convergence; 2000. Available from: http://thales.doa.fmph.uniba.sk/macaj/ICON.pdf [Accessed: 20 January 2020]

[15] Kostyrko P, Mačaj M, Šalát T, Strauch O. On statistical limit points. Proceedings of American Mathematical Society. 2001;**129**(9):2647-2654

[16] Šalát T. On statistically convergent sequences of real numbers. Mathematica Slovaca. 1980;**30**(2): 139-150

[17] Šalát T, Visnyai T. Subadditive measures on N and the convergence of series with positive terms. Acta Math. 2003;**6**:43-52

[18] Tóth JT, Filip F, Bukor J, Zsilinszky L. On $\mathcal{I}_{<q}$ and $\mathcal{I}_{\leq q}$ convergence of arithmetical functions. Periodica Mathematica Hungarica. 2020 [Accessed: 24 January 2020]

[19] Baláž V. Remarks on uniform density u. In: Proceedings IAM Workshop on Institute of Information Engineering, Automation and Mathematics. Bratislava: Slovak University of Technology; 2007. pp. 43-48

[20] Baláž V, Gogola J, Visnyai T. $\mathcal{I}_c^{(q)}$–convergence of arithmetical functions.

Journal of Number Theory. 2018;**183**: 74-83

[21] Fehér Z, László B, Mačaj M, Šalát T. Remarks on arithmetical functions $a_p(n), \gamma(n), \tau(n)$. Annals of Mathematics and Informaticae. 2006;**33**:35-43

[22] Furstenberg H. Recurrence in Ergodic Theory and Combinatorial Number Theory. Princeton: Princeton University Press; 1981

[23] Gogola J, Mačaj M, Visnyai T. On $\mathcal{I}_c^{(q)}$–convergence. Annals of Mathematics and Informaticae. 2011;**38**:27-36

[24] Gubo Š, Mačaj M, Šalát T, Tomanová J. On binomial coefficients. Acta Math. 2003;**6**:33-42

[25] Renling J. Applications of nonstandard analysis in additive number theory. The Bulletin of Symbolic Logic. 2000;**6**(3):331-341

[26] Šalát T. On the function $p^{a_p(n)} \| n \ (n > 1)$. Mathematica Slovaca. 1994;**44**(2):143-151

[27] Šalát T, Tomanová J. On the product of divisors of a positive integer. Mathematica Slovaca. 2002;**52**(3): 271-287

[28] Schinzel A, Šalát T. Remarks on maximum and minimum exponents in factoring. Mathematica Slovaca. 1994; **44**(5):505-514

[29] Petri H. Asymptotic properties of welfare relations. Economic Theory. 2019;**67**(4):853-874

[30] Ramsey FP. A mathematical theory of saving. The Economic Journal. 1928; **38**(152):543-559

[31] Fey M. May's theorem with an infinite population. Social Choice and Welfare. 2004;**23**:275-293

[32] Abbott H, Erdös P, Hanson D. On the number of times an integer occurs as a binomial coefficient. American Mathematical Monthly. 1974;**81**(3): 256-261

[33] Mycielski J. Sur les représentations des nombres naturels par des puissances à base et exposant naturels. Colloquium Mathematicum. 1951;**2**(3–4):254-260

[34] Hardy GH, Wright EM. An Introduction to the Theory of Numbers. 5th ed. Oxford: Clarendon Press; 1979

[35] Ostmann HH. Additive Zahlentheorie I. Berlin: Springer; 1956

[36] Paštéka M. On Four Approaches to Density. Bratislava: Peter Lang AG; 2014

[37] Paštéka M. Density and Related Topics. Nakladatelství Academia: Praha; 2017

[38] Strauch O, Porubský Š. Distribution of Sequences: A Sampler. Bern, Switzerland: Peter Lang; 2005

[39] Brown T, Freedman A. Arithmetic progressions in lacunary sets. Rocky Mountain Journal of Mathematics. 1987; **17**(3):587-596

[40] Freedman AR, Sember JJ. Densities and summability. Pacific Journal of Mathematics. 1981;**95**(2):293-305

[41] Kuratowski C. Topologie I. Warszawa: Panstwowe Wydawnictwa Naukowe; 1958

[42] Nagata JI. Modern General Topology. 2nd ed. Amsterdam: Elsevier; 1985

[43] Petersen GM. Regular Matrix Transformations. London: McGraw-Hill; 1966

[44] Miller HI. A measure theoretical subsequence characterization of statistical convergence. Transactions of

the American Mathematical Society. 1995;**347**(5):1811-1819

[45] Buck RC. The measure theoretic approach to density. American Journal of Mathematics. 1946;**68**(4):560-580

[46] Powell BJ, Šalát T. Convergence of subseries of the harmonic series and asymptotic densities of sets of positive integers. Publications de l'Institut Mathématique Nouvelle Série. 1991;**50**: 60-70

[47] Mikusiński P. Axiomatic theory of convergence. Pr Nauk Uniw ŚI Katow. 1982;**12**:13-21

[48] Mitrinovič DS, Sándor J, Crstici B. Handbook of Number Theory (Mathematics and its Applications). Vol. 351. Dordrecht: Kluwer Academic Publishers; 1995

Chapter 9

Identification of Eigen-Frequencies and Mode-Shapes of Beams with Continuous Distribution of Mass and Elasticity and for Various Conditions at Supports

Triantafyllos K. Makarios

Abstract

In the present article, an equivalent three degrees of freedom (DoF) system of two different cases of inverted pendulums is presented for each separated case. The first case of inverted pendulum refers to an amphi-hinge pendulum that possesses distributed mass and stiffness along its height, while the second case of inverted pendulum refers to an inverted pendulum with distributed mass and stiffness along its height. These vertical pendulums have infinity number of degree of freedoms. Based on the free vibration of the above-mentioned pendulums according to partial differential equation, a mathematically equivalent three-degree of freedom system is given for each case, where its equivalent mass matrix is analytically formulated with reference on specific mass locations along the pendulum height. Using the three DoF model, the first three fundamental frequencies of the real pendulum can be identified with very good accuracy. Furthermore, taking account the 3×3 mass matrix, it is possible to estimate the possible pendulum damages using a known technique of identification mode-shapes via records of response accelerations. Moreover, the way of instrumentation with a local network by three accelerometers is given via the above-mentioned three degrees of freedom.

Keywords: equivalent masses of continuous amphi-hinge vertical pendulum, identification of mode-shapes, distributed mass and stiffness, continuous systems, modal analysis of the continuous beam, inverted pendulum

1. Introduction

An ideal three degrees of freedom system that is equivalent with the modal behavior of an infinity number of degree of freedom of two cases of pendulums is proposed for each case. The first pendulum has hinges at the two ends, while the second pendulum is a cantilever (inverted pendulum). Both pendulums are presented analytically in the present article. This equivalent three DoF system can be used in instrumentation of such pendulums, where the concept of the concentrated masses is not existing, with a local network of three accelerometers. This issue is a main problem that appears very common during the instrumentation of inverted pendulums or bridge beams or steel stairs [1–3] or wind energy powers [4, 5] in order to identify the real vibration mode-shapes and the fundamental eigen-frequencies of the structure via records of response accelerograms at specific positions due to ambient excitation [6, 7].

2. First case: modal analysis of undamped amphi-hinge vertical pendulum with distributed mass and stiffness

According to the theory of continuous systems [8, 9], consider a straight amphi-hinge vertical pendulum that is loaded by an external continuous dynamic loading $p_z(x,t)$, with reference to a Cartesian three-dimensional reference system *oxyz*, (**Figure 1**).

The vertical pendulum possesses a distributed mass $m(x)$ per unit height, which in the special case of uniform distribution is given as $m(x) = \overline{m}$ in tons per meter (tn/m). Furthermore, according to Bernoulli Technical Bending Theory, the beam has section flexural stiffness $EI_y(x)$, where in the special case of an uniform distribution of the stiffness it is given as $EI_y(x) = EI_y$, where E is the material modulus of elasticity and I_y is the section moment of inertia about y-axis (**Figure 1**). Next, we are examining a such amphi-hinge vertical pendulum that possesses constant value of distributed mass along its height, as well as constant value of distributed section flexural stiffness EI_y. Due to fact that the vertical pendulum mass is continuously distributed, this pendulum/beam has infinity number of degrees of freedom for vibration along the horizontal *oz*-axis. In order to formulate of the motion equation

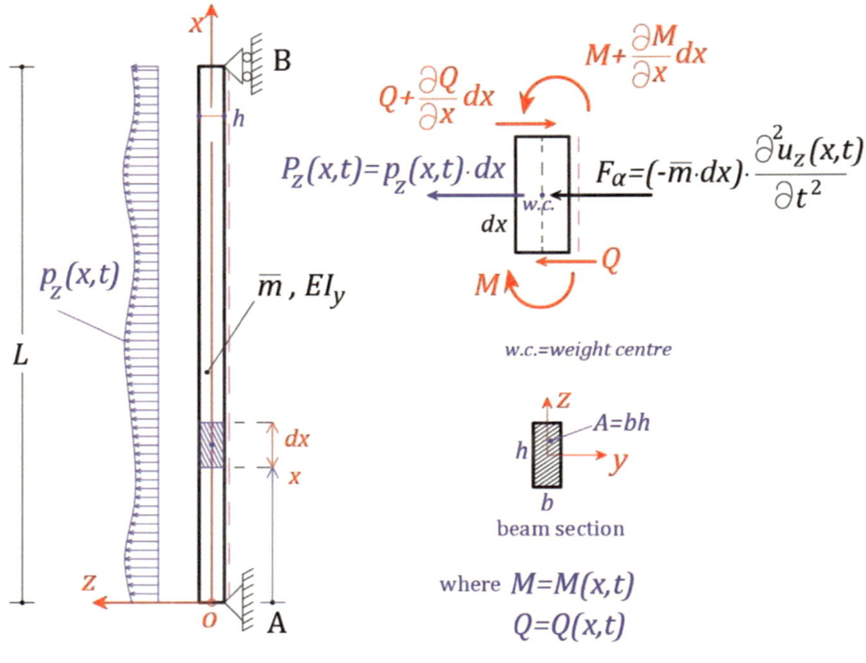

Figure 1.
Amphi-hinge vertical pendulum with distributed mass and section flexural stiffness.

of this beam, we consider an infinitesimal part of the vertical beam, at location x from the origin o, that has isolated by two very nearest parallel sections. The infinitesimal length of this part is the dx. On this infinitesimal length, we notice the flexural moment $M(x,t)$, the shear force $Q(x,t)$ with their differential increments, while the axial force $N(x,t)$ is ignored, because it does not affect the horizontal beam vibration along z-axis. Moreover, noted the resulting force $P_z(x,t)$ of the external dynamic loading. Therefore, we can write:

$$P_z(x,t) = p_z(x,t) \cdot dx \tag{1}$$

where the resulting force $P_z(x,t)$ acts at the total beam infinitesimal part.

Furthermore, according to D'Alembert principle, the resulting inertia force $F_a(x,t)$ is noted, where:

$$F_a(x,t) = (-\overline{m} \cdot dx) \cdot \partial^2 u_z(x,t)/\partial t^2 \Rightarrow F_a(x,t) = (-\overline{m} \cdot dx) \cdot \ddot{u}_z(x,t) \tag{2}$$

Here, we agree that the time derivatives of the displacements are going to symbolize with full stops, while the spatial derivatives of the displacements are going to symbolize with accent. Next, the damping and the second order differential are ignored, so the force equilibrium on the infinitesimal part of the beam along z-axis gives:

$$\sum F_z = 0 \ \Rightarrow Q + P_z(x,t) - \left(Q + \frac{\partial Q}{\partial x} dx\right) + F_a(x,t) = 0 \Rightarrow$$

$$\frac{\partial Q}{\partial x} = p_z(x,t) - \overline{m} \cdot \ddot{u}_z(x,t) \tag{3}$$

Moreover, the moment equilibrium with reference to weight center (w.c.) of the infinitesimal part of the beam (see **Figure 1**) gives:

$$\sum M_y = 0 \Rightarrow M + Q \cdot \frac{dx}{2} + \left(Q + \frac{\partial Q}{\partial x} dx\right) \cdot \frac{dx}{2} - \left(M + \frac{\partial M}{\partial x} dx\right) = 0 \Rightarrow$$

$$Q = \frac{\partial M}{\partial x} \tag{4}$$

According to Euler-Bernoulli Bending theory (where the shear deformations are ignored), it is well-known that the following basic equation is true:

$$M(x,t) = EI_y \cdot \frac{\partial^2 u_z(x,t)}{\partial x^2} \tag{5}$$

Eqs. (4) and (5) are inserted into Eq. (3), so the motion equation without damping for the examined vertical beam is given:

$$\frac{\partial^2 M}{\partial x^2} = p_z(x,t) - \overline{m} \cdot \ddot{u}_z(x,t) \Rightarrow$$

$$\frac{\partial^2}{\partial x^2}\left(EI_y \cdot \frac{\partial^2 u_z(x,t)}{\partial x^2}\right) = p_z(x,t) - \overline{m} \cdot \ddot{u}_z(x,t) \Rightarrow$$

$$\overline{m}\frac{\partial^2 u_z(x,t)}{\partial t^2} + EI_y \frac{\partial^4 u_z(x,t)}{\partial x^4} = p_z(x,t) \Rightarrow$$

$$\overline{m} \cdot \ddot{u}_z(x,t) + EI_y \cdot u_z''''(x,t) = p_z(x,t) \tag{6}$$

Eq. (6) is a partial differential equation that describes the motion $u_z(x,t)$ of the vertical beam that is loaded with the external dynamic loading $p_z(x,t)$. In order to arise a unique solution from Eq. (6), the support conditions must be used at the two beam ends. It is worthy to note that the classical case of a beam with distributed mass and section flexural stiffness, under external horizontal excitation (**Figure 2**) on the two supports is mathematically equivalent with the vibration that is described by Eq. (6). Indeed, in the case of **Figure 2**, the total displacement $u_z^{tot}(x,t)$ of the beam at x-location is given:

$$u_z^{tot}(x,t) = u_g(t) + u_z(x,t) \tag{7}$$

where $u_g(t)$ is the displacement at the base, same for the two supports.

But, it is known that the inertia forces of the beam are depended by the total displacement $u_z^{tot}(x,t)$, while the distributed dynamic loading is null, $p_z(x,t) = 0$. Thus, Eq. (3) is transformed into:

$$\frac{\partial Q}{\partial x} = p_z(x,t) - \overline{m} \cdot \frac{\partial^2 u_z^{tot}(x,t)}{\partial t^2} \Rightarrow$$

$$\frac{\partial Q}{\partial x} = 0 - \overline{m} \cdot \frac{\partial^2 u_g(t)}{\partial t^2} - \overline{m} \cdot \frac{\partial^2 u_z(x,t)}{\partial t^2} \tag{8}$$

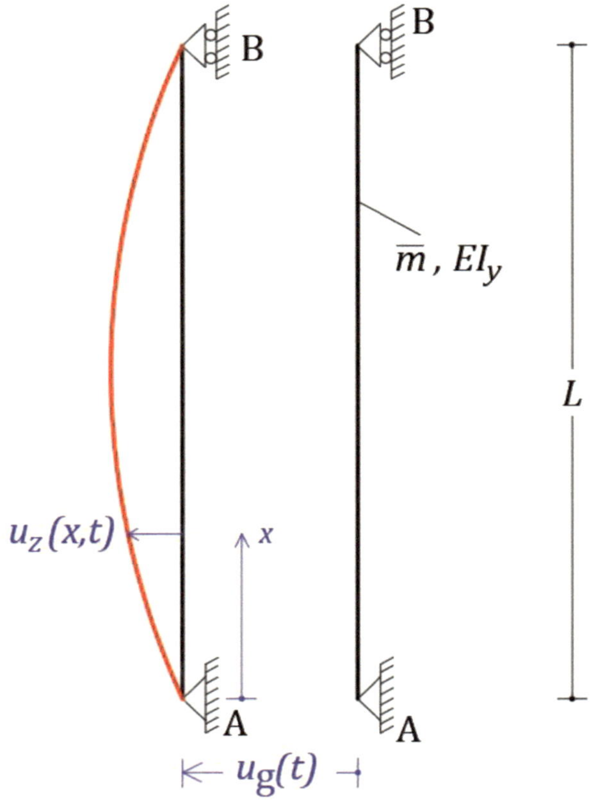

Figure 2.
Amphi-hinge vertical pendulum subjected with the same horizontal ground motion $u_g(t)$ on the two end-supports.

Following, Eqs. (4) and (5) are inserting into Eq. (8), thus we are taken:

$$\frac{\partial^2 M}{\partial x^2} = -\overline{m} \cdot \left(\frac{\partial^2 u_g(t)}{\partial t^2} + \frac{\partial^2 u_z(x,t)}{\partial t^2} \right) \Rightarrow$$

$$\overline{m} \frac{\partial^2 u_z(x,t)}{\partial t^2} + EI_y \frac{\partial^4 u_z(x,t)}{\partial x^4} = -\overline{m} \cdot \frac{\partial^2 u_g(t)}{\partial t^2} \tag{9}$$

By the comparison of Eqs. (6) and (9), we notice that the undamped beam vibration due to horizontal motion of the two supports is mathematically equivalent with the undamped vibration of the same beam where the two supports are fixed and the beam is loaded with the equivalent distributed dynamic loading $p_{eq}(x,t)$:

$$p_{eq}(x,t) = -\overline{m} \cdot \frac{\partial^2 u_g(t)}{\partial t^2} \tag{10}$$

In the case of the horizontal pendulum/beam free vibration without damping, we consider the first part of Eq. (9) that must be null:

$$\overline{m} \frac{\partial^2 u_z(x,t)}{\partial t^2} + EI_y \frac{\partial^4 u_z(x,t)}{\partial x^4} = 0 \tag{11}$$

Furthermore, we ask the unknown spatial time-function $u_z(x,t)$, which is the solution of Eq. (11), must have the form of separated variants:

$$u_z(x,t) = \varphi(x) \cdot q(t) \tag{12}$$

where $\varphi(x)$ is an unknown spatial function and $q(t)$ is an unknown time-function. Eq. (12) has been derived two times with reference to time-dimension t and more two times with reference to spatial-dimension x, so:

$$\frac{\partial^2 u_z(x,t)}{\partial t^2} = \varphi(x) \cdot \ddot{q}(t), \quad \frac{\partial^2 u_z(x,t)}{\partial x^2} = \varphi''(x) \cdot q(t) \tag{13}$$

Eqs. (13) are inserted into Eq. (11), giving:

$$\overline{m} \cdot \varphi(x) \cdot \ddot{q}(t) + EI_y \cdot \varphi''''(x) \cdot q(t) = 0 \tag{14}$$

and, next, divided with the number $\overline{m} \cdot \varphi(x) \cdot q(t)$, thus we are getting:

$$\frac{-\ddot{q}(t)}{q(t)} = \frac{EI_y \cdot \varphi''''(x)}{\overline{m} \cdot \varphi(x)} \tag{15}$$

The left part of Eq. (15) is a time-function, but the right part is a spatial-function. In order to true Eq. (15) for all time values as well as for all spatial positions, the two parts of Eq. (15) must be equal with a constant λ. Thus, Eq. (15) is separated at two following differential equations:

$$\frac{-\ddot{q}(t)}{q(t)} = \lambda \Rightarrow \ddot{q}(t) + \lambda \cdot q(t) = 0 \tag{16}$$

$$\frac{EI_y \cdot \varphi''''(x)}{\overline{m} \cdot \varphi(x)} = \lambda \Rightarrow EI_y \cdot \varphi^{''''(x)} - \lambda \cdot \overline{m} \cdot \varphi(x) = 0 \tag{17}$$

However, the time equation (16) indicates a free vibration of an ideal single degree of freedom system that has eigen-frequency $= \sqrt{\lambda}$. Inserting the eigen-frequency ω into Eq. (17) gives:

$$EI_y \cdot \varphi'''' (x) - \omega^2 \cdot \overline{m} \cdot \varphi(x) = 0 \Rightarrow \varphi''''(x) - \frac{\omega^2 \cdot \overline{m}}{EI_y} \cdot \varphi(x) = 0 \tag{18}$$

Next, we set the positive parameter β such as to be equal:

$$\beta^4 = \frac{\omega^2 \cdot \overline{m}}{EI_y} \tag{19}$$

because the parameters $\omega^2, \overline{m}, EI_y$ are always positive. By the mathematic theory it is known that the general solution of Eq. (18) has the following form:

$$\varphi(x) = C_1 \sin \beta x + C_2 \cos \beta x + C_3 \sinh \beta x + C_4 \cosh \beta x \tag{20}$$

where the four unknown parameters C_1, C_2, C_3, C_4 must be calculated. In order to achieve this, four support conditions of the beam have to used. Indeed, for $x = 0$ and $x = L$ the displacement $u_z(0,t)$ of the amphi-hinge vertical pendulum as well as the flexural moment $M(0,t)$, both are equal zero. The spatial function $\varphi(x)$, which is the solution of Eq. (20), gives the modal elastic line of the beam. Having as known data that the following equation is true:

$$\sinh \beta x = \frac{e^{\beta x} - e^{-\beta x}}{2}, \ \cosh \beta x = \frac{e^{\beta x} + e^{-\beta x}}{2} \tag{21}$$

The spatial function of the modal elastic line for $x = 0$ is:

$$\varphi(0) = C_1 \sin 0 + C_2 \cos 0 + C_3 \sinh 0 + C_4 \cosh 0 = 0 \Rightarrow$$
$$C_2 + C_4 = 0 \tag{22}$$

and also for $x = 0$, the function of the flexural moment due to examined modal elastic line of the beam is given by Eq. (5):

$$M(0,t) = EI_y \cdot \frac{\partial^2 \varphi(0)}{\partial x^2} = 0 \Rightarrow EI_y \cdot \varphi''(0) = 0 \tag{23}$$

Eq. (20) has been derived two times with reference to spatial-dimension x, thus arise:

$$\varphi' (x) = C_1 \cdot \beta \cdot \cos \beta x + C_2 \cdot (-\beta) \cdot \sin \beta x + C_3 \cdot \beta \cdot \cosh \beta x + C_4 \cdot \beta \cdot \sinh \beta x \tag{24}$$

and

$$\varphi''(x) = C_1(-\beta^2) \cdot \sin \beta x + C_2(-\beta^2) \cdot \cos \beta x + C_3\beta^2 \cdot \sinh \beta x + C_4\beta^2 \cdot \cosh \beta x \tag{25}$$

Therefore, Eq. (23) is transformed:

$$EI_y \cdot \beta^2 \cdot (C_4 - C_2) = 0 \tag{26}$$

By Eqs. (22) and (26) directly arise $C_2 = 0$ and $C_4 = 0$, thus the general solution of Eq. (20) is the following:

$$\varphi(x) = C_1 \sin\beta x + C_3 \sinh\beta x \tag{27}$$

In addition, the parameters C_1, C_3 are calculated by the support conditions of the second support of the beam. Therefore, for $x = L$ the vertical displacement $u(L,t) = 0$ be true. Thus, from Eq. (12) arises that $\varphi(L) = 0$ and Eq. (27) gives:

$$\varphi(L) = C_1 \sin\beta L + C_3 \sinh\beta L = 0 \tag{28}$$

In continuous, Eq. (23) gives:

$$M(L,t) = EI_y \cdot \frac{\partial^2 \varphi(L)}{\partial x^2} = 0 \Rightarrow EI_y \cdot \varphi''(L) = 0 \tag{29}$$

where $\varphi''(L)$ is directly getting from Eq. (25) that is equivalent with zero:

$$\varphi''(L) = C_1(-\beta^2) \cdot \sin\beta L + C_3\beta^2 \cdot \sinh\beta L = 0 \tag{30}$$

However, when Eqs. (28) and (30) are re-written again, we get:

$$C_1 \cdot \sin\beta L + C_3 \cdot \sinh\beta L = 0 \tag{31}$$

$$-C_1 \cdot \sin\beta L + C_3 \cdot \sinh\beta L = 0 \tag{32}$$

And added part to part these two above-mentioned equations arise:

$$2 \cdot C_3 \cdot \sinh\beta L = 0 \tag{33}$$

But, the term $\sinh\beta L$ is not equal with zero, because then vibration is not existing. Therefore, C_3 has to equal with zero, so Eq. (28) is formed:

$$\varphi(L) = C_1 \cdot \sin\beta L = 0 \tag{34}$$

Moreover, by Eq. (34) arise that either $C_1 = 0$ that is mpossibility because $\varphi(x) \neq 0$ by Eq. (27), either $\sin\beta L = 0$ that means the following equation must be true:

$$\beta L = n \cdot \pi \qquad n = 1, 2, 3, \ldots \tag{35}$$

However, Eq. (35) is transformed to Eq. (36):

$$\beta L = n \cdot \pi \Rightarrow \beta^2 L^2 = n^2 \cdot \pi^2 \Rightarrow \beta^2 = \frac{n^2 \cdot \pi^2}{L^2} \tag{36}$$

By the definition of parameter β, we can calculate the eigen-frequency ω:

$$\beta^4 = \frac{\omega^2 \cdot \overline{m}}{EI_y} \Rightarrow \omega^2 = \frac{EI_y \cdot \beta^4}{\overline{m}} \Rightarrow \omega = \beta^2 \cdot \sqrt{\frac{EI_y}{\overline{m}}} \tag{37}$$

Thus, inserting Eq. (36) into Eq. (37), the eigen-frequency ω_n directly arises for each n-value.

$$\omega_n = \frac{n^2 \cdot \pi^2}{L^2} \cdot \sqrt{\frac{EI_y}{\overline{m}}} \qquad n = 1, 2, 3, \ldots \tag{38}$$

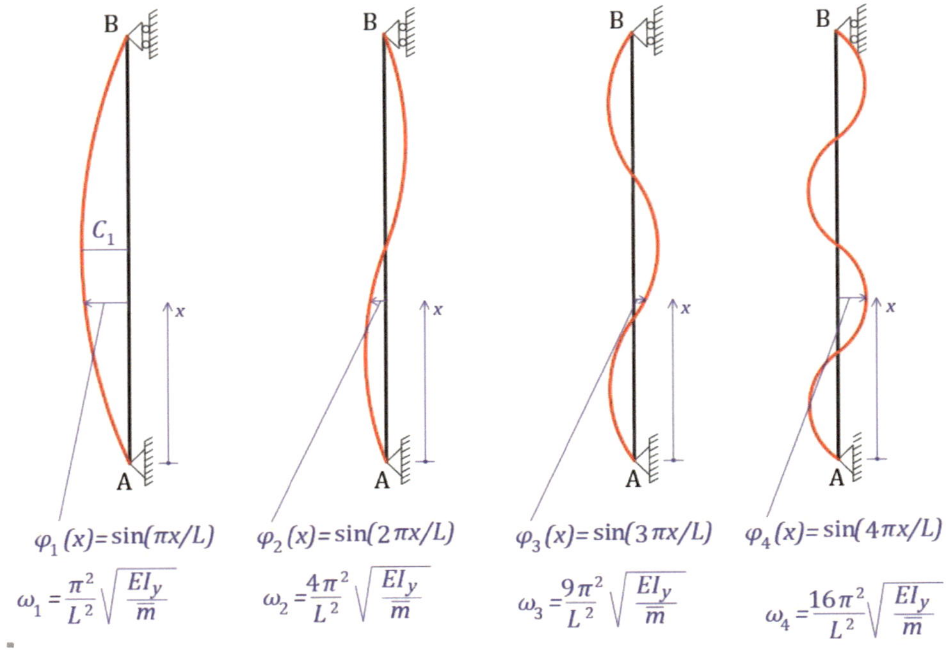

Figure 3.
Modal analysis of continuous amphi-hinge vertical pendulum. The first four mode-shapes.

Therefore, the vibration mode-shape of the examined vertical pendulum arises by Eq. (27)—since previous inserting Eq. (35)—thus:

$$\varphi_n(x) = C_1 \sin \beta x = C_1 \sin \frac{n \cdot \pi \cdot x}{L} \qquad n = 1, 2, 3, \ldots \tag{39}$$

The value of C_1 is arbitrary, and we usually get it equal to unit. Thus, for each value of parameter n, a mode-shape with its eigen-frequency is resulted. The fundamental (first) mode-shape is resulted for $n = 1$, which shows a half sinusoidal wave, the second mode-shape shows a full sinusoidal wave, etc. (**Figure 3**). The order of the eigen-frequencies is ω_1, $\omega_2 = 4\omega_1$, $\omega_3 = 9\omega_1$, $\omega_4 = 16\omega_1$, etc.

3. The equivalent three degrees of freedom system for amphi-hinges pendulum

At vertical pendulums or beams where the fundamental horizontal mode-shape does not activate the 90% of the total beam mass, we ask to consider the three first mode-shapes. Thus, for this purpose, we must define an ideal equivalent three degrees of freedom beam, which is going to give the first three frequencies and the first three mode-shapes of the examined beam. Therefore, which is the ideal three degrees of freedom system, where its three mode-shapes coincide with the real first three frequencies of the vertical pendulum with distributed mass and flexural stiffness?

In order to answer the above-mentioned question, consider a weightless vertical pendulum with height L and constant section in elevation, where carry three concentrated masses that each one has the same mass-value m_{eq}, located per distance $0.25L$, between one to one, and each one mass possesses an horizontal degree of freedom (**Figure 4**).

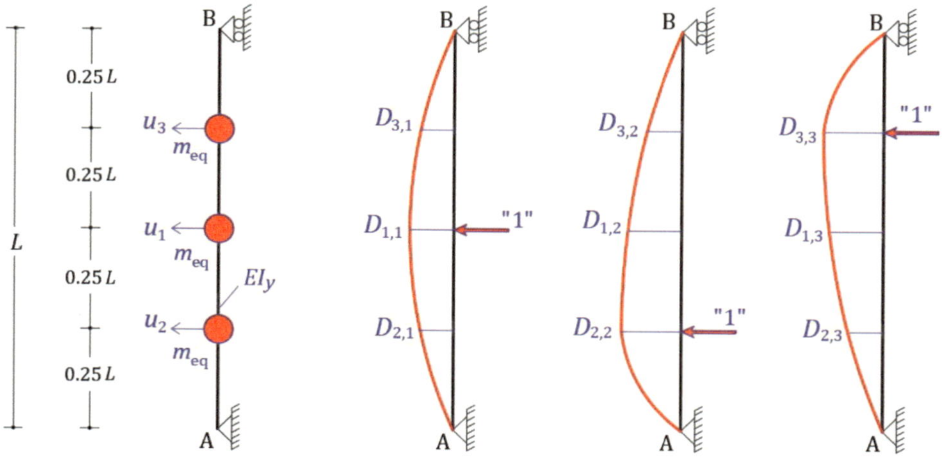

Figure 4.
The equivalent three-degree of freedom system on amphi-hinge vertical pendulum.

The pendulum displacement vector **u** of the three degrees of freedom, as well as the diagonal beam mass matrix $\boldsymbol{m}$ is written:

$$\mathbf{u} = \begin{Bmatrix} u_1 \\ u_2 \\ u_3 \end{Bmatrix}, \mathbf{m} = \begin{bmatrix} m_{eq} & 0 & 0 \\ 0 & m_{eq} & 0 \\ 0 & 0 & m_{eq} \end{bmatrix} \tag{40}$$

Furthermore, the pendulum flexibility matrix $\boldsymbol{f}$ can be calculated using a suitable method (**Figure 4**), and the inverse matrix gives the stiffness matrix $\boldsymbol{k}$ of the three degrees of freedom beam.

$$\boldsymbol{f} = \begin{bmatrix} D_{1,1} & D_{1,2} & D_{1,3} \\ D_{2,1} & D_{2,2} & D_{2,3} \\ D_{3,1} & D_{3,2} & D_{3,3} \end{bmatrix} = \frac{L^3}{48EI_y} \cdot \begin{bmatrix} 1 & 0.6875 & 0.6875 \\ 0.6875 & 0.5625 & 0.4375 \\ 0.6875 & 0.4375 & 0.5625 \end{bmatrix} \tag{41}$$

$$\boldsymbol{k} = \begin{bmatrix} k_{1,1} & k_{1,2} & k_{1,3} \\ k_{2,1} & k_{2,2} & k_{2,3} \\ k_{3,1} & k_{3,2} & k_{3,3} \end{bmatrix} = \frac{48EI_y}{L^3} \cdot \begin{bmatrix} 18.285714 & -12.571429 & -12.571429 \\ -12.571429 & 13.142857 & 5.142857 \\ -12.571429 & 5.142857 & 13.142857 \end{bmatrix} \tag{42}$$

The equations of motion for the case of the free undamped vibration of the ideal beam is given as:

$$\mathbf{m}\,\ddot{\mathbf{u}}(t) + \mathbf{k}\,\mathbf{u}(t) = \mathbf{0} \tag{43}$$

The eigen-problem is written as:

$$\left(\mathbf{k} - \omega_n^2\mathbf{m}\right)\boldsymbol{\varphi}_n = \mathbf{0} \qquad n = 1, 2, 3. \tag{44}$$

where the eigen-frequencies $\boldsymbol{\omega}_n$ and the three mode-shapes $\boldsymbol{\varphi}_n$ are known by Eq. (38) and (39) and **Figure 4**. Therefore, the unique unknown parameter is the mass m_{eq}. Thus,

$$\det\left(\mathbf{k}-\omega_1^2\mathbf{m}\right)=0\Rightarrow \tag{45}$$

$$m_{\text{eq}}^3+A\cdot m_{\text{eq}}^2+B\cdot m_{\text{eq}}+C=0 \tag{46}$$

where

$$A=-\frac{k_{11}+k_{22}+k_{33}}{\omega_1^2},B=\frac{k_{11}k_{33}+k_{11}k_{22}+k_{22}k_{33}-k_{12}^2-k_{13}^2-k_{23}^2}{\omega_1^4}$$

$$C=-\frac{k_{11}k_{22}k_{33}+2k_{12}k_{13}k_{23}-k_{11}k_{23}^2-k_{22}k_{13}^2-k_{33}k_{12}^2}{\omega_1^6}$$

The numerical solution of Eq. (46) gives three roots for parameter m_{eq}, where only the first root is acceptable, because the other two values rejected since do not have natural meaning (appear values greater from the total pendulum mass $\overline{m}L$). Thus, the only one acceptable root is given as:

$$m_{\text{eq}}=0.24984748\cdot\left(\overline{m}L\right) \tag{47}$$

Therefore, inserting the ideal equivalent mass m_{eq} by Eq. (47) at three degrees of freedom system of **Figure 4**, the three eigen-frequencies coincide with the real values of the initial vertical pendulum that has distributed mass and flexural stiffness.

4. Discussion about the amphi-hinge vertical pendulums

By the previous mathematic analysis, an ideal three degrees of freedom system that is equivalent with the modal behavior of the amphi-hinge vertical pendulum with distributed mass and flexural stiffness along its height has been presented. This ideal three degrees of freedom system can be used in instrumentation of a such vertical tower, which does not possess concentrated masses. In the framework of the identification of mode-shapes of an amphi-hinge pendulum, the equivalent mass by Eq. (47) permits to locate accelerometers per 0.25L (as shown at **Figure 4**) and there measure the response horizontal acceleration histories, in order to calculate the real first three mode-shapes of the vertical pendulum in order to avoid the instability and resonance-vibrations between the examined vertical pendulum and the supported rockets or space bus before the last launched.

5. Second case: modal analysis of undamped inverted pendulum with distributed mass and stiffness

According to the theory of continuous systems [8, 9], consider a straight vertical cantilever (inverted pendulum) that is loaded by an external continuous dynamic loading $p_z(x,t)$, with reference to a Cartesian three-dimensional reference system $oxyz$, (**Figure 5**).

The vertical inverted pendulum possesses a distributed mass $m(x)$ per unit height, which in the special case of uniform distribution is given as $m(x)=\overline{m}$ in tons per meter (tn/m). Furthermore, according to Bernoulli Technical Bending Theory, this cantilever has section flexural stiffness $EI_y(x)$, where in the special case of an uniform distribution of the stiffness it is given as $EI_y(x)=EI_y$, where E is the

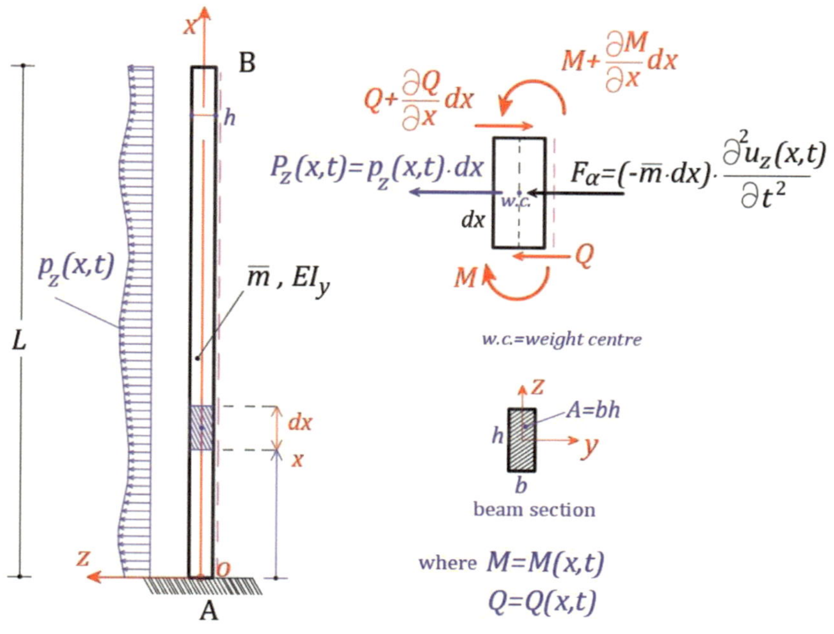

Figure 5.
Vertical cantilever (inverted pendulum) with distributed mass and section flexural stiffness.

material modulus of elasticity and I_y is the section moment of inertia about y-axis (**Figure 1**). Next, we are examining such an inverted pendulum that possesses constant value of distributed mass along its height, as well as constant value of distributed section flexural stiffness EI_y. Due to fact that the cantilever mass is continuously distributed, this inverted pendulum/beam has infinity number of degrees of freedom for vibration along the horizontal oz-axis. In order to formulate of the motion equation of this beam, we consider an infinitesimal part of the vertical beam, at location x from the origin o, that has isolated by two very nearest parallel sections.

The infinitesimal length of this part is the dx. On this infinitesimal length, we notice the flexural moment $M(x,t)$, the shear force $Q(x,t)$ with their differential increments, while the axial force $N(x,t)$ is ignored, because it does not affect the horizontal cantilever vibration along z-axis. Moreover, noted the resulting force $P_z(x,t)$ of the external dynamic loading. Eqs. (1)–(20) describe mathematically the modal analysis of the inverted pendulum. Afterward, the four unknown parameters C_1, C_2, C_3, C_4 of Eq. (20) must be calculated taking account the conditions of the cantilever. In order to achieve this, four support conditions of the cantilever have to be used. Indeed, for the end at $x = 0$ the displacement $u_z(0,t)$ and the slope $u'_z(0,t)$ of displacement profile must be zero. So, Eq. (20) gives:

$$\varphi(0) = C_1 \sin 0 + C_2 \cos 0 + C_3 \sinh 0 + C_4 \cosh 0 = 0 \Rightarrow C_2 + C_4 = 0 \qquad (48)$$

and

$$\varphi'(0) = \beta(C_1 + C_3) = 0 \Rightarrow C_1 + C_3 = 0 \qquad (49)$$

Moreover, for the free end at $x = L$ of the cantilever, the flexural moment $M(L,t)$ as well as the shear force $Q(L,t)$ must be both zero. So, from Eqs. (20) and (5) and afterward from Eq. (48) we take:

$$M(L,t) = EI_y \cdot \frac{\partial^2 \varphi(L)}{\partial x^2} = 0 \Rightarrow EI_y \cdot \varphi''(L) = 0 \Rightarrow$$
$$C_1(\sin \beta L + \sinh \beta L) + C_2(\cos \beta L + \cosh \beta L) = 0 \tag{50}$$

and

$$Q(L,t) = EI_y \cdot \frac{\partial^3 \varphi(L)}{\partial x^3} = 0 \Rightarrow EI_y \cdot \varphi'''(L) = 0 \Rightarrow$$
$$C_1(\cos \beta L + \cosh \beta L) + C_2(-\sin \beta L + \sinh \beta L) = 0 \tag{51}$$

However, re-writing Eqs. (50) and (51) again, we get the matrix form:

$$\begin{bmatrix} (\sin \beta L + \sinh \beta L) & (\cos \beta L + \cosh \beta L) \\ (\cos \beta L + \cosh \beta L) & (-\sin \beta L + \sinh \beta L) \end{bmatrix} \begin{Bmatrix} C_1 \\ C_2 \end{Bmatrix} = \begin{Bmatrix} 0 \\ 0 \end{Bmatrix} \tag{52}$$

Eq. (52) is a real eigenvalue problem. In order to calculate the eigenvalues, parameters C_1 and C_2 must not both equal zero. Thus, the determinant of the matrix by Eq. (52) must be zero, where it drives to the following frequency equation:

$$1 + (\cos \beta L) \cdot (\cosh \beta L) = 0 \tag{53}$$

However, Eq. (54) can be solved numerically only, where we obtain the first four roots ($n = 1, 2, 3, 4$):

$$\beta_1 L = 1.8751, \quad \beta_2 L = 4.6941, \quad \beta_3 L = 7.8548 \text{ and } \quad \beta_4 L = 10.996 \tag{54}$$

By the definition of parameter β, we can calculate the first four circular eigen-frequency ω using Eq. (36):

$$\beta^4 = \frac{\omega^2 \cdot \overline{m}}{EI_y} \Rightarrow \omega^2 = \frac{EI_y \cdot \beta^4}{\overline{m}} \Rightarrow \omega = \beta^2 \cdot \sqrt{\frac{EI_y}{\overline{m}}} \tag{55}$$

Thus, inserting Eq. (54) into Eq. (55), the first four eigen-frequency ω_n is directly arise for each n-value.

$$\omega_1 = \frac{3.516}{L^2} \cdot \sqrt{\frac{EI_y}{\overline{m}}}, \quad \omega_2 = \frac{22.03}{L^2} \cdot \sqrt{\frac{EI_y}{\overline{m}}},$$
$$\omega_3 = \frac{61.70}{L^2} \cdot \sqrt{\frac{EI_y}{\overline{m}}}, \quad \omega_4 = \frac{120.9}{L^2} \cdot \sqrt{\frac{EI_y}{\overline{m}}} \tag{56}$$

Therefore, the vibration mode-shape of the examined vertical inverted pendulum arises by Eq. (20)—since previous inserting Eq. (56)—thus:

$$\varphi_n(x) = C_1 \left[\cosh \beta_n x - \cos \beta_n x - \frac{\cosh \beta_n L + \cos \beta_n L}{\sinh \beta_n L + \sin \beta_n L} (\sinh \beta_n x - \sin \beta_n x) \right] \tag{57}$$

The value of C_1 is an arbitrary constant, and we usually get it equal to unit. Thus, for each value of parameter n, a mode-shape with its eigen-frequency is resulted. The fundamental (first) mode-shape is resulted for $n = 1$, etc. (**Figure 6**).

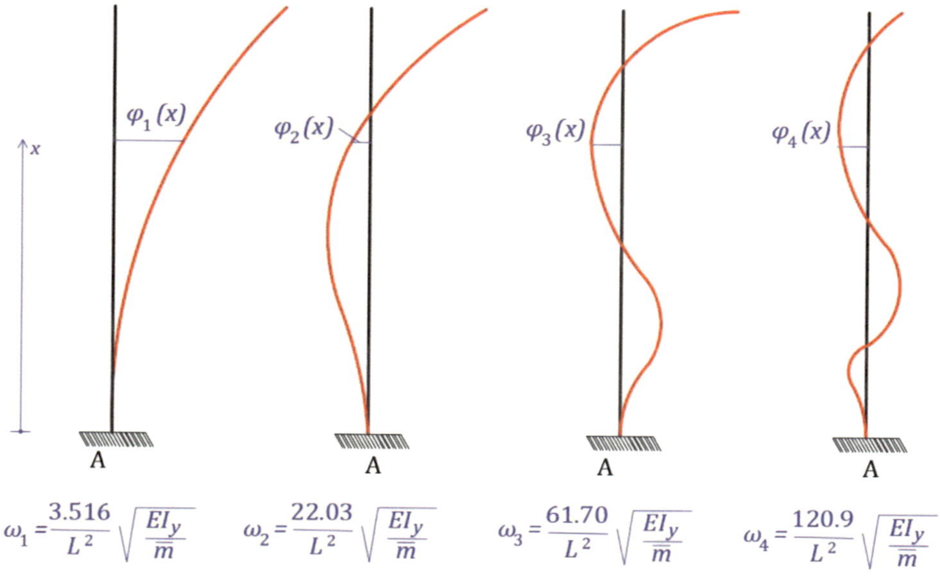

Figure 6.
Modal analysis of continuous inverted pendulum. The first four mode-shapes.

6. The equivalent three degrees of freedom system of inverted pendulum

At inverted pendulums cantilevers, where the fundamental horizontal mode-shape does not activate the 90% of the total cantilever mass, we ask to consider the three first mode-shapes. Thus, for this purpose, we must define an ideal equivalent three degrees of freedom beam, which is going to give the three mode-shapes of the examined beam. Therefore, which is the ideal three degrees of freedom system, where its three eigen-frequencies coincide with the real first three frequencies of the inverted pendulum with distributed mass and flexural stiffness?

In order to answer the above-mentioned question, consider a weightless vertical inverted pendulum with height L and constant section in elevation, where carry three concentrated masses that each one has the same mass-value m_{eq}, located per distance $0.333L$, between one to one, and each one mass possesses an horizontal degree of freedom (**Figure 7**).

The inverted pendulum displacement vector **u** of the three degrees of freedom, as well as the diagonal beam mass matrix ***m*** are written:

$$\mathbf{u} = \begin{Bmatrix} u_1 \\ u_2 \\ u_3 \end{Bmatrix}, \mathbf{m} = \begin{bmatrix} m_{eq} & 0 & 0 \\ 0 & m_{eq} & 0 \\ 0 & 0 & m_{eq} \end{bmatrix} \tag{58}$$

Furthermore, the inverted pendulum flexibility matrix f can be calculated using a suitable method (**Figure 7**), and the inverse matrix gives the stiffness matrix $\boldsymbol{k}$ of the three degrees of freedom beam.

$$f = \begin{bmatrix} D_{1,1} & D_{1,2} & D_{1,3} \\ D_{2,1} & D_{2,2} & D_{2,3} \\ D_{3,1} & D_{3,2} & D_{3,3} \end{bmatrix} = \frac{L^3}{3EI_y} \cdot \begin{bmatrix} 1 & 0.5185 & 0.1481 \\ 0.5185 & 0.2963 & 0.0926 \\ 0.1481 & 0.0926 & 0.0370 \end{bmatrix} \tag{59}$$

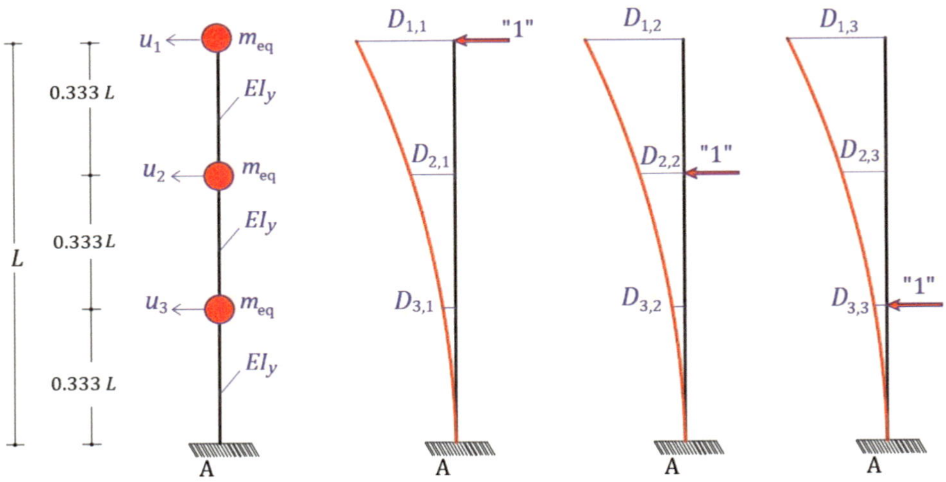

Figure 7.
The equivalent three-degree of freedom system on inverted pendulum.

$$\boldsymbol{k} = \begin{bmatrix} k_{1,1} & k_{1,2} & k_{1,3} \\ k_{2,1} & k_{2,2} & k_{2,3} \\ k_{3,1} & k_{3,2} & k_{3,3} \end{bmatrix} = \frac{3EI_y}{L^3} \cdot \begin{bmatrix} 14.53846 & -33.23077 & 24.92308 \\ -33.23077 & 91.38462 & -95.53846 \\ 24.92308 & -95.53846 & 166.15385 \end{bmatrix} \tag{60}$$

The equations of motion for the case of the free undamped vibration of the ideal beam is given as:

$$\mathbf{m}\,\ddot{\mathbf{u}}(t) + \mathbf{k}\,\mathbf{u}(t) = \mathbf{0} \tag{61}$$

The eigen-problem is written as:

$$(\mathbf{k} - \omega_n^2\mathbf{m})\boldsymbol{\varphi}_n = \mathbf{0} \quad n = 1, 2, 3. \tag{62}$$

where the eigen-frequencies $\boldsymbol{\omega}_n$ and the three mode-shapes $\boldsymbol{\varphi}_n$ are known by Eq. (56) and **Figure 6**. Therefore, the unique unknown parameter is the mass m_{eq}. Thus,

$$\det(\mathbf{k} - \omega_1^2\mathbf{m}) = 0 \Rightarrow \tag{63}$$

$$m_{\text{eq}}^3 + A \cdot m_{\text{eq}}^2 + B \cdot m_{\text{eq}} + C = 0 \tag{64}$$

where

$$A = -\frac{k_{11} + k_{22} + k_{33}}{\omega_1^2}, B = \frac{k_{11}k_{33} + k_{11}k_{22} + k_{22}k_{33} - k_{12}^2 - k_{13}^2 - k_{23}^2}{\omega_1^4}$$

$$C = -\frac{k_{11}k_{22}k_{33} + 2k_{12}k_{13}k_{23} - k_{11}k_{23}^2 - k_{22}k_{13}^2 - k_{33}k_{12}^2}{\omega_1^6}$$

The numerical solution of Eq. (64) gives three roots for parameter m_{eq}, where only the first root is acceptable, because the other two values rejected since do not have natural meaning (appear values greater from the total pendulum mass $\overline{m}L$). Thus, the only one acceptable root is given:

$$m_{\text{eq}} = 0.1868388 \cdot (\overline{m}L) \tag{65}$$

Therefore, inserting the ideal equivalent mass m_{eq} by Eq. (65) at three degrees of freedom system of **Figure 7**, the three eigen-frequencies coincide with the real values of the initial vertical inverted pendulum that has distributed mass and flexural stiffness.

On the contrary to above-mentioned about the examined cantilever, it is worth noting that in the case where ask an ideal single degree of freedom system that has eigen-frequency equal to fundamental eigen-frequency of the real cantilever we write the following equations:

- Eigen-frequency of single degree of freedom system with k the cantilever lateral stiffness and $m_{\text{eq, sdof}}$ the concentrated mass at the top of the cantilever:

$$\omega^2 = \frac{k}{m_{\text{eq,sdof}}} = \frac{3EI_y/L^3}{m_{\text{eq,sdof}}} \tag{66}$$

- Fundamental (first) eigen-frequency (see Eq. (56)) of the real cantilever with continuous distribution of mass and flexural stiffness:

$$\omega = \frac{3.516}{L^2} \cdot \sqrt{\frac{EI_y}{\overline{m}}} \tag{67}$$

- Thus, inserting Eq. (67) into Eq. (66) the equivalent concentrated mass $m_{\text{eq,sdof}}$ at the top of the cantilever is given [9]:

$$m_{\text{eq,sdof}} = 0.2426742 \cdot (\overline{m}L) \tag{68}$$

7. Conclusions

The present article has presented a mathematic ideal three degrees of freedom system that is equivalent with the modal behavior of two cases of pendulums. First, an amphi-hinge vertical pendulum with distributed mass and flexural stiffness along its height has been examined. Second, an inverted pendulum that can be simulates a tower (or cantilever) has been examined too. For each case, an ideal three degrees of freedom system has been proposed that can be used in instrumentation of such a vertical tower, which does not possess concentrated masses. In the framework of the identification of mode-shapes of the above-mentioned pendulums, the equivalent mass by Eqs. (47) and (65) permits to locate accelerometers (as shown at **Figures 4** and 7) and measures the response acceleration histories, in order to calculate the real first three frequencies.

Acknowledgements

The author wishes to acknowledge the German Research Foundation (DFG) program on Initiation of International Collaboration entitled "Data-driven analysis models for slender structures using explainable artificial intelligence", Project No. 417973400 for the period 2019–2021, Prof. Dr.-Ing. Kay Smarsly, Bauhaus University Weimar, project coordinator.

Author details

Triantafyllos K. Makarios
Institute of Structural Analysis and Dynamics of Structures, School of Civil Engineering, Faculty of Engineer, Aristotle University of Thessaloniki, Thessaloniki Town, Greece

*Address all correspondence to: makariostr@civil.auth.gr

References

[1] Manolis GD, Makarios TK, Terzi V, Karetsou I. Mode shape identification of an existing three-story flexible steel stairway as a continuous dynamic system. Theoretical and Applied Mechanics. 2015;**42**(3):151-166

[2] Makarios TK, Manolis G, Karetsou I, Papanikolaou M, Terzi V. Modelling and identification of the dynamic response of an existing three-story steel stairway. In: COMPDYN 2015, 5th ECCOMAS Thematic Conference on Computational Methods in Structural Dynamics and Earthquake Engineering; 25–27 May. Crete Island, Greece; 2015

[3] Makarios T, Manolis G, Terzi V, Karetsou I. Identification of dynamic characteristics of a continuous system: Case study for a flexible steel stairway. In: 16th World Conference on Earthquake; 9–13 January; 16WCEE 2017 (paper 1011). Santiago Chile; 2017

[4] Makarios T, Baniotopoulos C. Wind energy structures: Modal analysis by the continuous model approach. Journal of Vibration and Control. 2014;**20**(3): 395-405

[5] Makarios T, Baniotopoulos C. Modal analysis of wind turbine tower via its continuous model with partially fixed foundation. International Journal of Innovative Research in Advanced Engineering. 2015;**2**(1):14-25

[6] Makarios T. Identification of the mode shapes of spatial tall multi-storey buildings due to earthquakes. The new "modal time-histories" method. Journal of the Structural Design of Tall & Special Buildings. 2012;**21**(9):621-641

[7] Makarios T. Chapter 4: Identification of building dynamic characteristics by using the modal response acceleration time-histories in the seismic excitation and the wind dynamic loading cases. In: Accelerometers; Principles, Structure and Applications. New York: Nova Science Publisher; 2013. pp. 77-113. ISBN: 978-1-62808-128-2

[8] Chopra A. Dynamics of Structures. Theory and Applications to Earthquake Engineering. International Edition. Englewood Cliffs, New Jersey: Prentice-Hall, Inc.; 1995. p. 07632

[9] Clough R, Penzien J. Dynamics of Structures. Third edition. Berkeley, USA: McGraw-Hill; 1995

Chapter 10

Some Identities Involving 2-Variable Modified Degenerate Hermite Polynomials Arising from Differential Equations and Distribution of Their Zeros

Cheon Seoung Ryoo

Abstract

In this chapter, we introduce the 2-variable modified degenerate Hermite polynomials and obtain some new symmetric identities for 2-variable modified degenerate Hermite polynomials. In order to give explicit identities for 2-variable modified degenerate Hermite polynomials, differential equations arising from the generating functions of 2-variable modified degenerate Hermite polynomials are studied. Finally, we investigate the structure and symmetry of the zeros of the 2-variable modified degenerate Hermite equations.

Keywords: differential equations, symmetric identities, modified degenerate Hermite polynomials, complex zeros

1. Introduction

The Hermite equation is defined as

$$u''(x) - 2xu'(x) + (\varepsilon - 1)u(x) = 0, x \in [-\infty, \infty], \tag{1}$$

where ε is unrestricted. Hermite equation is encountered in the study of quantum mechanical harmonic oscillator, where ε represent the energy of the oscillator. The ordinary Hermite numbers H_n and Hermite polynomials $H_n(x)$ are usually defined by the generating functions

$$e^{t(2x-t)} = \sum_{n=0}^{\infty} H_n(x)\frac{t^n}{n!} \tag{2}$$

and

$$e^{-t^2} = \sum_{n=0}^{\infty} H_n \frac{t^n}{n!}. \tag{3}$$

Clearly, $H_n = H_n(0)$.

It is known that these numbers and polynomials play an important role in various fields of mathematics and physics, including number theory, combinations, special functions, and differential equations. Many interested properties about that have been studied (see [1–5]). The ordinary Hermite polynomials $H_n(x)$ satisfy the Hermite differential equation

$$\frac{d^2H(x)}{dx^2} - 2x\frac{dH(x)}{dx} + 2nH(x) = 0, n = 0, 1, 2, \ldots . \tag{4}$$

Hence ordinary Hermite polynomials $H_n(x)$ satisfy the second-order ordinary differential equation

$$u'' - 2xu' + 2nu = 0. \tag{5}$$

We remind that the 2-variable Hermite polynomials $H_n(x,y)$ defined by the generating function (see [2])

$$\sum_{n=0}^{\infty} H_n(x,y)\frac{t^n}{n!} = e^{t(x+yt)} \tag{6}$$

are the solution of heat equation

$$\frac{\partial}{\partial y}H_n(x,y) = \frac{\partial^2}{\partial x^2}H_n(x,y), \quad H_n(x,0) = x^n. \tag{7}$$

Observe that

$$H_n(2x,-1) = H_n(x). \tag{8}$$

Motivated by their importance and potential applications in certain problems of probability, combinatorics, number theory, differential equations, numerical analysis and other areas of mathematics and physics, several kinds of some special numbers and polynomials were recently studied by many authors (see [1–8]). Many mathematicians have studied in the area of the degenerate Stiling, degenerate Bernoulli polynomials, degenerate Euler polynomials, degenerate Genocchi polynomials, and degenerate tangent polynomials (see [6, 7, 9]).

Recently, Hwang and Ryoo [10] proposed the 2-variable degenerate Hermite polynomials $\mathcal{H}_n(x,y,\lambda)$ by means of the generating function

$$\sum_{n=0}^{\infty} \mathcal{H}_n(x,y,\lambda)\frac{t^n}{n!} = (1+\lambda)^{\frac{t(x+yt)}{\lambda}}. \tag{9}$$

Since $(1+\mu)^{\frac{t}{\mu}} \to e^t$ as $\mu \to 0$, it is evident that (9) reduces to (6). The 2-variable degenerate Hermite polynomials $\mathcal{H}_n(x,y,\lambda)$ in generating function (9) are the solution of equation

$$\begin{aligned} &\frac{\partial}{\partial y}\mathcal{H}_n(x,y,\lambda) = \frac{\lambda}{\log(1+\lambda)}\frac{\partial^2}{\partial x^2}\mathcal{H}_n(x,y,\lambda), \\ &\mathcal{H}_n(x,0,\lambda) = \left(\frac{\log(1+\lambda)}{\lambda}\right)^n x^n. \end{aligned} \tag{10}$$

Since $\frac{\log(1+\lambda)}{\lambda} \to 1$ as λ approaches to 0, it is apparent that (10) descends to (7).

Mathematicians have studied the differential equations arising from the generating functions of special numbers and polynomials (see [10–14]). Now, a new class of 2-variable modified degenerate Hermite polynomials are constructed based on the results so far. We can induce the differential equations generated from the generating function of 2-variable modified degenerate Hermite polynomials. By using the coefficients of this differential equation, we obtain explicit identities for the 2-variable modified degenerate Hermite polynomials. The rest of the paper is organized as follows. In Section 2, we construct the 2-variable modified degenerate Hermite polynomials and obtain basic properties of these polynomials. In Section 3, we give some symmetric identities for 2-variable modified degenerate Hermite polynomials. In Section 4, we derive the differential equations generated from the generating function of 2-variable modified degenerate Hermite polynomials. Using the coefficients of this differential equation, we have explicit identities for the 2-variable modified degenerate Hermite polynomials. In Section 5, we investigate the zeros of the 2-variable modified degenerate Hermite equations by using computer. Further, we observe the pattern of scattering phenomenon for the zeros of 2-variable modified degenerate Hermite equations. Our paper will finish with Section 6, where the conclusions and future directions of this work are showed.

2. Basic properties for the 2-variable modified degenerate Hermite polynomials

In this section, a new class of the 2-variable modified degenerate Hermite polynomials are considered. Furthermore, some properties of these polynomials are also obtained.

We define the 2-variable modified degenerate Hermite polynomials $\mathrm{H}_n(x,y|\mu)$ by means of the generating function

$$\sum_{n=0}^{\infty} \mathrm{H}_n(x,y|\mu)\frac{t^n}{n!} = (1+\mu)^{\frac{xt}{\mu}} e^{yt^2}. \tag{11}$$

Since $(1+\mu)^{\frac{xt}{\mu}} \to e^{xt}$ as $\mu \to 0$, it is clear that (11) reduces to (6). Observe that degenerate Hermite polynomials $\mathcal{H}_n(x,y,\mu)$ and 2-variable modified degenerate Hermite polynomials $\mathrm{H}_n(x,y|\mu)$ are totally different.

Now, we recall that the μ-analogue of the falling factorial sequences as follows:

$$(x|\mu)_0 = 1, (x|\mu)_n = x(x-\mu)(x-2\mu)\cdots(x-(n-1)\mu), (n \geq 1). \tag{12}$$

Note that $\lim_{\mu\to 1}(x|\mu)_n = x(x-1)(x-2)\cdots(x-(n-1)) = (x)_n, (n \geq 1)$. We also need the binomial theorem: for a variable y,

$$\begin{aligned}(1+\mu)^{yt/\mu} &= \sum_{m=0}^{\infty} \binom{ty}{\mu}_m \frac{\mu^m}{m!} \\ &= \sum_{m=0}^{\infty} \left(\sum_{l=0}^{m} S_1(m,l) \left(\frac{ty}{\mu}\right)^l \frac{\mu^m}{m!} \right) \\ &= \sum_{l=0}^{\infty} \left(\sum_{m=l}^{\infty} S_1(m,l) y^l \mu^{m-l} \frac{l!}{m!} \right) \frac{t^l}{l!}. \end{aligned} \tag{13}$$

We remember that the classical Stirling numbers of the first kind $S_1(n,k)$ and the second kind $S_2(n,k)$ are defined by the relations (see [6–13])

$$(x)_n = \sum_{k=0}^{n} S_1(n,k)x^k \text{ and } x^n = \sum_{k=0}^{n} S_2(n,k)(x)_k, \tag{14}$$

respectively. We also have

$$\sum_{n=m}^{\infty} S_2(n,m)\frac{t^n}{n!} = \frac{(e^t-1)^m}{m!} \text{ and } \sum_{n=m}^{\infty} S_1(n,m)\frac{t^n}{n!} = \frac{(\log(1+t))^m}{m!}. \tag{15}$$

As another application of the differential equation for $\mathrm{H}_n(x,y|\mu)$ is as follows: Note that

$$G(t,x,y,\mu) = (1+\mu)^{\frac{xt}{\mu}} e^{yt^2} \tag{16}$$

satisfies

$$\frac{\partial G(t,x,y,\mu)}{\partial y} - \left(\frac{\log(1+\mu)}{\mu}\right)^2 \frac{\partial^2 G(t,x,y,\mu)}{\partial x^2} = 0. \tag{17}$$

Substitute the series in (11) for $G(t,x,y,\mu)$ to get

$$\frac{\partial}{\partial y}\mathrm{H}_n(x,y|\mu) = \left(\frac{\mu}{\log(1+\mu)}\right)^2 \frac{\partial^2}{\partial x^2}\mathrm{H}_n(x,y|\mu).$$

Thus the 2-variable modified degenerate Hermite polynomials $\mathrm{H}_n(x,y|\mu)$ in generating function (11) are the solution of equation

$$\begin{aligned} &\left(\frac{\log(1+\mu)}{\mu}\right)^2 \frac{\partial}{\partial y}\mathrm{H}_n(x,y|\mu) - \frac{\partial^2}{\partial x^2}\mathrm{H}_n(x,y|\mu) = 0, \\ &\mathrm{H}_n(x,0|\mu) = \left(\frac{\log(1+\mu)}{\mu}\right)^n x^n. \end{aligned} \tag{18}$$

The generating function (11) is useful for deriving several properties of the 2-variable modified degenerate Hermite polynomials $\mathrm{H}_n(x,y|\mu)$. For example, we have the following expression for these polynomials:

Theorem 1. For any positive integer n, we have

$$\mathrm{H}_n(x,y|\mu) = \sum_{k=0}^{\left[\frac{n}{2}\right]} \left(\frac{\log(1+\mu)}{\mu}\right)^{n-2k} x^{n-2k} y^k \frac{n!}{k!(n-2k)!}, \tag{19}$$

where [ ] denotes taking the integer part.

Proof. By (11) and (13), we have

$$\begin{aligned} \sum_{n=0}^{\infty} \mathrm{H}_n(x,y|\mu)\frac{t^n}{n!} &= (1+\mu)^{\frac{xt}{\mu}} e^{yt^2} \\ &= \sum_{k=0}^{\infty} y^k \frac{t^{2k}}{k!} \sum_{l=0}^{\infty} \left(\frac{\log(1+\mu)}{\mu}\right)^l x^l \frac{t^l}{l!} \\ &= \sum_{n=0}^{\infty} \left(\sum_{k=0}^{\left[\frac{n}{2}\right]} y^k \left(\frac{\log(1+\mu)}{\mu}\right)^{n-2k} x^{n-2k} \frac{n!}{k!(n-2k)!} \right) \frac{t^n}{n!}. \end{aligned} \tag{20}$$

By comparing the coefficients of $\frac{t^n}{n!}$, the expected result of Theorem 1 is achieved.

□

Since $\lim_{\mu\to 0}\frac{\log(1+\mu)}{\mu}=1$, we get

$$H_n(x,y)=n!\sum_{k=0}^{\left[\frac{n}{2}\right]}\frac{y^k x^{n-2k}}{k!(n-2k)!}. \tag{21}$$

The following basic properties of the 2-variable degenerate Hermite polynomials $\mathrm{H}_n(x,y|\mu)$ are induced form (11). Therefore, it is enough to delete involved detail explanation.

Theorem 2. For any positive integer n, we have

$$\begin{aligned}
&1.\ \mathrm{H}_n(x,y|\mu)=\sum_{k=0}^{\left[\frac{n}{2}\right]}\sum_{m=n-2k}^{\infty}y^k S_1(m,n-2k)x^{n-2k}\mu^{m-(n-2k)}\frac{n!}{m!k!}.\\
&2.\ \mathrm{H}_n(x_1+x_2,y|\mu)=\sum_{l=0}^{n}\binom{n}{l}\left(\frac{\log(1+\mu)}{\mu}\right)^l x_2^l\mathrm{H}_{n-l}(x_1,y|\mu).\\
&3.\ \mathrm{H}_n(x_1+x_2,y|\mu)=\sum_{l=0}^{n}\binom{n}{l}\mathrm{H}_{n-l}(x_1,y|\mu)\sum_{m=l}^{\infty}S_1(m,l)x_2^l\mu^{m-l}\frac{l!}{m!}.\\
&4.\ \mathrm{H}_n(x,y_1+y_2|\mu)=\sum_{k=0}^{\left[\frac{n}{2}\right]}\mathrm{H}_{n-2k}(x,y_1|\mu)\frac{y_2^k n!}{k!(n-2k)!}.\\
&5.\ \mathrm{H}_n(x_1+x_2,y_1+y_2|\mu)=\sum_{l=0}^{n}\binom{n}{l}\mathrm{H}_l(x_1,y_1|\mu)\mathrm{H}_{n-l}(x_2,y_2|\mu).
\end{aligned} \tag{22}$$

3. Symmetric identities for 2-variable modified degenerate Hermite polynomials

In this section, we give some new symmetric identities for 2-variable modified degenerate Hermite polynomials. We also get some explicit formulas and properties for 2-variable modified degenerate Hermite polynomials.

Theorem 3. Let $a,b>0$ $(a\neq b)$. The following identity holds true:

$$b^m\mathrm{H}_m(ax,a^2y|\mu)=a^m\mathrm{H}_m(bx,b^2y|\mu). \tag{23}$$

Proof. Let $a,b>0$ $(a\neq b)$. We start with

$$\mathcal{G}(t,\mu)=(1+\mu)^{\frac{abxt}{\mu}}e^{a^2b^2yt^2}. \tag{24}$$

Then the expression for $\mathcal{G}(t,\mu)$ is symmetric in a and b

$$\mathcal{G}(t,\mu)=\sum_{m=0}^{\infty}\mathrm{H}_m(ax,a^2y|\mu)\frac{(bt)^m}{m!}=\sum_{m=0}^{\infty}b^m\mathrm{H}_m(ax,a^2y|\mu)\frac{t^m}{m!}. \tag{25}$$

By the similar way, we get that

$$\mathcal{G}(t,\mu)=\sum_{m=0}^{\infty}\mathrm{H}_m(bx,b^2y|\mu)\frac{(at)^m}{m!}=\sum_{m=0}^{\infty}a^m\mathrm{H}_m(bx,b^2y|\mu)\frac{t^m}{m!}. \tag{26}$$

By comparing the coefficients of $\frac{t^m}{m!}$ in last two equations, the expected result of Theorem 3 is achieved. □

Again, we now use

$$\mathcal{F}(t,\mu) = \frac{abt(1+\mu)^{\frac{abxt}{\mu}} e^{a^2b^2yt^2}\left((1+\mu)^{\frac{abt}{\mu}} - 1\right)}{\left((1+\mu)^{\frac{at}{\mu}} - 1\right)\left((1+\mu)^{\frac{bt}{\mu}} - 1\right)}. \tag{27}$$

For $\mu \in \mathbb{C}$, we introduce the modified degenerate Bernoulli polynomials given by the generating function

$$\sum_{n=0}^{\infty} \beta_n(x|\mu)\frac{t^n}{n!} = \frac{t}{(1+\mu)^{\frac{t}{\mu}} - 1}(1+\mu)^{\frac{xt}{\mu}}, \text{(see[6, 7])}. \tag{28}$$

When $x = 0$ and $\beta_n(\mu) = \beta_n(0|\mu)$ are called the modified degenerate Bernoulli numbers. Note that

$$\lim_{\mu \to 0} \beta_n(\mu) = B_n, \tag{29}$$

where B_n are called the Bernoulli numbers. The first few of them are

$$\begin{aligned}
\beta_0(x|\mu) &= \frac{\mu}{\log(1+\mu)}, \\
\beta_1(x|\mu) &= -\frac{1}{2} + x, \\
\beta_2(x|\mu) &= \frac{\log(1+\mu)}{6\mu} - \frac{x\log(1+\mu)}{\mu} + \frac{x^2\log(1+\mu)}{\mu}, \\
\beta_3(x|\mu) &= \frac{x\log(1+\mu)^2}{2\mu^2} - \frac{3x^2\log(1+\mu)^2}{2\mu^2} + \frac{x^3\log(1+\mu)^2}{\mu^2}, \\
\beta_4(x|\mu) &= -\frac{\log(1+\mu)^3}{30\mu^3} + \frac{x^2\log(1+\mu)^3}{\mu^3} - \frac{2x^3\log(1+\mu)^3}{\mu^3} + \frac{x^4\log(1+\mu)^3}{\mu^3}.
\end{aligned} \tag{30}$$

For each integer $k \geq 0$, $S_k(n) = 0^k + 1^k + 2^k + \cdots + (n-1)^k$ is called sum of integers. A modified generalized falling factorial sum $\sigma_k(n,\mu)$ can be defined by the generating function

$$\sum_{k=0}^{\infty} \sigma_k(n|\mu)\frac{t^k}{k!} = \frac{(1+\mu)^{\frac{(n+1)t}{\mu}} - 1}{(1+\mu)^{\frac{t}{\mu}} - 1}. \tag{31}$$

Note that $\lim_{\mu\to 0}\sigma_k(n|\mu) = S_k(n)$. From $\mathcal{F}(t,\mu)$, we get the following result:

$$\begin{aligned}
\mathcal{F}(t,\mu) &= \frac{abt(1+\mu)^{\frac{abxt}{\mu}} e^{a^2b^2yt^2}\left((1+\mu)^{\frac{abt}{\mu}} - 1\right)}{\left((1+\mu)^{\frac{at}{\mu}} - 1\right)\left((1+\mu)^{\frac{bt}{\mu}} - 1\right)} \\
&= \frac{abt}{\left((1+\mu)^{\frac{at}{\mu}} - 1\right)}(1+\mu)^{\frac{abxt}{\mu}} e^{a^2b^2yt^2}\frac{\left((1+\mu)^{\frac{abt}{\mu}} - 1\right)}{\left((1+\mu)^{\frac{bt}{\mu}} - 1\right)} \\
&= b\sum_{n=0}^{\infty}\beta_n(\mu)\frac{(at)^n}{n!}\sum_{n=0}^{\infty}\mathrm{H}_n(bx, b^2y|\mu)\frac{(at)^n}{n!}\sum_{n=0}^{\infty}\sigma_k(a-1|\mu)\frac{(bt)^n}{n!} \\
&= \sum_{n=0}^{\infty}\left(\sum_{i=0}^{n}\sum_{m=0}^{i}\binom{n}{i}\binom{i}{m}a^i b^{n+1-i}\beta_m(\mu)\mathrm{H}_{i-m}(bx, b^2y|\mu)\sigma_{n-i}(a-1|\mu)\right)\frac{t^n}{n!}.
\end{aligned} \tag{32}$$

In a similar fashion we have

$$
\begin{aligned}
\mathcal{F}(t,\mu) &= \frac{abt}{\left((1+\mu)^{\frac{bt}{\mu}}-1\right)}(1+\mu)^{\frac{abxt}{\mu}}e^{a^2b^2yt^2}\frac{\left((1+\mu)^{\frac{abt}{\mu}}-1\right)}{\left((1+\mu)^{\frac{at}{\mu}}-1\right)} \\
&= a\sum_{n=0}^{\infty}\beta_n(\mu)\frac{(bt)^n}{n!}\sum_{n=0}^{\infty}\mathrm{H}_n\left(ax,a^2y|\mu\right)\frac{(bt)^n}{n!}\sum_{n=0}^{\infty}\sigma_k(b-1|\mu)\frac{(at)^n}{n!} \\
&= \sum_{n=0}^{\infty}\left(\sum_{i=0}^{n}\sum_{m=0}^{i}\binom{n}{i}\binom{i}{m}b^ia^{n+1-i}\beta_m(\mu)\mathrm{H}_{i-m}\left(ax,a^2y|\mu\right)\sigma_{n-i}(b-1|\mu)\right)\frac{t^n}{n!}.
\end{aligned}
\tag{33}
$$

By comparing the coefficients of $\frac{t^m}{m!}$ on the right hand sides of the last two equations, we have the below theorem.

Theorem 4. Let $a,b>0(a\neq b)$. The the following identity holds true:

$$
\begin{aligned}
&\sum_{i=0}^{n}\sum_{m=0}^{i}\binom{n}{i}\binom{i}{m}a^ib^{n+1-i}\beta_m(\mu)\mathrm{H}_{i-m}\left(bx,b^2y|\mu\right)\sigma_{n-i}(a-1|\mu) \\
&=\sum_{i=0}^{n}\sum_{m=0}^{i}\binom{n}{i}\binom{i}{m}b^ia^{n+1-i}\beta_m(\mu)\mathrm{H}_{i-m}\left(ax,a^2y|\mu\right)\sigma_{n-i}(b-1|\mu).
\end{aligned}
\tag{34}
$$

By taking the limit as $\mu\to 0$, we have the following corollary.

Corollary 5. Let $a,b>0$ $(a\neq b)$. The the following identity holds true:

$$
\begin{aligned}
&\sum_{i=0}^{n}\sum_{m=0}^{i}\binom{n}{i}\binom{i}{m}a^ib^{n+1-i}\mathrm{S}_{n-i}(a-1)B_mH_{i-m}\left(bx,b^2y\right) \\
&=\sum_{i=0}^{n}\sum_{m=0}^{i}\binom{n}{i}\binom{i}{m}b^ia^{n+1-i}\mathrm{S}_{n-i}(b-1)B_mH_{i-m}\left(ax,a^2y\right).
\end{aligned}
\tag{35}
$$

4. Differential equations associated with 2-variable modified degenerate Hermite polynomials

In this section, we construct the differential equations with coefficients $a_i(N,x,y,\mu)$ arising from the generating functions of the 2-variable modified degenerate Hermite polynomials:

$$
\begin{aligned}
&\left(\frac{\partial}{\partial t}\right)^N G(t,x,y,\mu)-a_0(N,x,y,\mu)G(t,x,y,\mu)-\cdots \\
&-a_{2N}(N,x,y,\mu)t^NG(t,x,y,\mu)=0.
\end{aligned}
\tag{36}
$$

By using the coefficients of this differential equation, we can get explicit identities for the 2-variable modified degenerate Hermite polynomials $\mathrm{H}_n(x,y,\mu)$. Recall that

$$
\begin{aligned}
G &= G(t,x,y,\mu) \\
&= (1+\mu)^{\frac{xt}{\mu}}e^{yt^2} \\
&= \sum_{n=0}^{\infty}\mathrm{H}_n(x,y|\mu)\frac{t^n}{n!},\quad \mu,x,t\in\mathbb{C}.
\end{aligned}
\tag{37}
$$

Then, by (37), we have

$$\begin{aligned} G^{(1)} &= \frac{\partial}{\partial t} G(t,x,y,\mu) \\ &= \frac{\partial}{\partial t}\left((1+\mu)^{\frac{xt}{\mu}} e^{yt^2}\right) \\ &= \left(\frac{\log(1+\mu)}{\mu}x + 2yt\right)\left((1+\mu)^{\frac{xt}{\mu}} e^{yt^2}\right) \\ &= \left(\frac{\log(1+\mu)}{\mu}x + 2yt\right) G(t,x,y,\mu), \end{aligned} \tag{38}$$

$$\begin{aligned} G^{(2)} &= \frac{\partial}{\partial t} G^{(1)}(t,x,y,\mu) \\ &= 2yG(t,x,y,\mu) + \left(\frac{\log(1+\mu)}{\mu}x + 2yt\right) G^{(1)}(t,x,y,\mu) \\ &= \left(2y + \left(\frac{\log(1+\mu)}{\mu}\right)^2 x^2\right) G(t,x,y,\mu) \\ &+ \left(\frac{\log(1+\mu)}{\mu} 4xy\right) tG(t,x,y,\mu) \\ &+ (2y)^2 t^2 G(t,x,y,\mu). \end{aligned} \tag{39}$$

By continuing this process as shown in (39), we can get easily that

$$\begin{aligned} G^{(N)} &= \left(\frac{\partial}{\partial t}\right)^N G(t,x,y,\mu) \\ &= \sum_{i=0}^{N} a_i(N,x,y,\mu) t^i G(t,x,y,\mu), \ (N = 0, 1, 2, \ldots). \end{aligned} \tag{40}$$

By differentiating (40) with respect to t, we have

$$\begin{aligned} G^{(N+1)} &= \frac{\partial G^{(N)}}{\partial t} = \sum_{i=0}^{N} a_i(N,x,y,\mu)(i) t^{i-1} G(t,x,y,\mu) \\ &\quad + \sum_{i=0}^{N} a_i(N,x,y,\mu) t^i G^{(1)}(t,x,y,\mu) \\ &= \sum_{i=0}^{N} (i) a_i(N,x,y,\mu) t^{i-1} G(t,x,y,\mu) \\ &\quad + \sum_{i=0}^{N} \left(\frac{\log(1+\mu)}{\mu} x\right) a_i(N,x,y,\mu) t^i G(t,x,y,\mu) \\ &\quad + \sum_{i=0}^{N} (2y) a_i(N,x,y,\mu) t^{i+1} G(t,x,y,\mu) \\ &= \sum_{i=0}^{N-1} (i+1) a_{i+1}(N,x,y,\mu) t^i G(t,x,y,\mu) \\ &\quad + \sum_{i=0}^{N} \left(\frac{\log(1+\mu)}{\mu} x\right) a_i(N,x,y,\mu) t^i G(t,x,y,\mu) \\ &\quad + \sum_{i=1}^{N+1} (2y) a_{i-1}(N,x,y,\mu) t^i G(t,x,y,\mu). \end{aligned} \tag{41}$$

Now we replace N by $N+1$ in (40). We find

$$G^{(N+1)} = \sum_{i=0}^{N+1} a_i(N+1,x,y,\mu)t^i G(t,x,y,\mu). \tag{42}$$

By comparing the coefficients on both sides of (41) and (42), we get

$$\begin{aligned} a_0(N+1,x,y,\mu) &= a_1(N,x,y,\mu) + \left(\frac{\log(1+\mu)}{\mu}x\right)a_0(N,x,y,\mu), \\ a_N(N+1,x,y,\mu) &= \left(\frac{\log(1+\mu)}{\mu}x\right)a_N(N,x,y,\mu) \\ &\quad +2ya_{N-1}(N,x,y,\mu), \\ a_{N+1}(N+1,x,y,\mu) &= (2y)a_N(N,x,y,\mu), \end{aligned} \tag{43}$$

and

$$\begin{aligned} a_i(N+1,x,y,\mu) &= (i+1)a_{i+1}(N,x,y,\mu) \\ &\quad + \left(\frac{\log(1+\mu)}{\mu}x\right)a_i(N,x,y,\mu) \\ &\quad +(2y)a_{i-1}(N,x,y,\mu), (1 \le i \le N-1). \end{aligned} \tag{44}$$

In addition, by (37), we have

$$G(t,x,y,\mu) = G^{(0)}(t,x,y,\mu) = a_0(0,x,y,\mu)G(t,x,y,\mu). \tag{45}$$

By (45), we get

$$a_0(0,x,y,\mu) = 1. \tag{46}$$

It is not difficult to show that

$$\begin{aligned} &\frac{x\log(1+\mu)}{\mu}G(t,x,y,\mu) + 2ytG(t,x,y,\mu) \\ &= G^{(1)}(t,x,y,\mu) \\ &= \sum_{i=0}^{1} a_i(1,x,y,\mu)t^i G(t,x,y,\mu) \\ &= a_0(1,x,y,\mu)G(t,x,y,\mu) + a_1(1,x,y,\mu)tG(t,x,y,\mu). \end{aligned} \tag{47}$$

Thus, by (38) and (47), we also get

$$a_0(1,x,y,\mu) = \frac{x\log(1+\mu)}{\mu}, \quad a_1(1,x,y,\mu) = 2y. \tag{48}$$

From (43) and (44), we note that

$$\begin{aligned} a_0(N+1,x,y,\mu) &= a_1(N,x,y,\mu) + \frac{x\log(1+\mu)}{\mu}a_0(N,x,y,\mu), \\ a_0(N,x,y,\mu) &= a_1(N-1,x,y,\mu) + \frac{x\log(1+\mu)}{\mu}a_0(N-1,x,y,\mu), \\ &\cdots \\ a_0(N+1,x,y,\mu) &= \sum_{i=0}^{N}\left(\frac{x\log(1+\mu)}{\mu}\right)^i a_1(N-i,x,y,\mu) \\ &\quad +\left(\frac{\log(1+\mu)}{\mu}\right)^{N+1}x^{N+1}, \end{aligned} \tag{49}$$

$$\begin{aligned} a_N(N+1,x,y,\mu) &= \frac{x\log(1+\mu)}{\mu}a_N(N,x,y,\mu) \\ &\quad +(2y)a_{N-1}(N,x,y,\mu), \\ a_{N-1}(N,x,y,\mu) &= \frac{x\log(1+\mu)}{\mu}a_{N-1}(N-1,x,y,\mu) \\ &\quad +(2y)a_{N-2}(N-1,x,y,\mu), \dots \\ a_N(N+1,x,y,\mu) &= (N+1)x(2y)^N\left(\frac{\log(1+\mu)}{\mu}\right), \end{aligned} \tag{50}$$

and

$$\begin{aligned} a_{N+1}(N+1,x,y,\mu) &= (2y)a_N(N,x,y,\mu), \\ a_N(N,x,y,\mu) &= (2y)a_{N-1}(N-1,x,y,\mu), \dots \\ a_{N+1}(N+1,x,y,\mu) &= (2y)^{N+1}. \end{aligned} \tag{51}$$

For $i=1$ in (44), we have

$$\begin{aligned} a_1(N+1,x,y,\mu) &= 2\sum_{k=0}^{N}\left(\frac{x\log(1+\mu)}{\mu}\right)^k a_2(N-k,x,y,\mu) \\ &\quad +2y\sum_{k=0}^{N}\left(\frac{x\log(1+\mu)}{\mu}\right)^k a_0(N-k,x,y,\mu), \end{aligned} \tag{52}$$

Continuing this process, we can deduce that, for $1 \le i \le N-1$,

$$\begin{aligned} a_i(N+1,x,y,\mu) &= (i+1)\sum_{k=0}^{N}\left(\frac{x\log(1+\mu)}{\mu}\right)^k a_{i+1}(N-k,x,y,\mu) \\ &\quad +2y\sum_{k=0}^{N}\left(\frac{x\log(1+\mu)}{\mu}\right)^k a_{i-1}(N-k,x,y,\mu). \end{aligned} \tag{53}$$

Note that, from (37)–(53), here the matrix $a_i(j,x,y,\mu)_{0\le i,j\le N+1}$ is given by

$$\begin{pmatrix} 1 & \frac{x\log(1+\mu)}{\mu} & 2y+\left(\frac{\log(1+\mu)}{\mu}\right)^2 x^2 & \cdots & \cdot \\ 0 & 2y & \left(\frac{\log(1+\mu)}{\mu}\right)4xy & \cdots & \cdot \\ 0 & 0 & (2y)^2 & \cdots & \cdot \\ \vdots & \vdots & \vdots & \ddots & \cdot \\ 0 & 0 & 0 & \cdots & (2y)^{N+1} \end{pmatrix} \tag{54}$$

Therefore, from (37)–(53), we obtain the following theorem.

Theorem 5. For $N=0,1,2,\dots,$ the differential equation

$$\left(\frac{\partial}{\partial t}\right)^N G(t,x,y,\mu) - \left(\sum_{i=0}^{N} a_i(N,x,y,\mu)t^i\right)G(t,x,y,\mu) = 0 \tag{55}$$

has a solution

$$G = G(t,x,y,\mu) = (1+\mu)^{\frac{xt}{\mu}}e^{yt^2}, \tag{56}$$

where

$$\begin{aligned}
a_0(N+1,x,y,\mu) &= \sum_{i=0}^{N}\left(\frac{x\log(1+\mu)}{\mu}\right)^i a_1(N-i,x,y,\mu) \\
&\quad+\left(\frac{\log(1+\mu)}{\mu}\right)^{N+1}x^{N+1}, \\
a_N(N+1,x,y,\mu) &= (N+1)(2y)^N\left(\frac{x\log(1+\mu)}{\mu}\right), \\
a_{N+1}(N+1,x,y,\mu) &= (2y)^{N+1}, \\
a_i(N+1,x,y,\mu) &= (i+1)\sum_{k=0}^{N}\left(\frac{x\log(1+\mu)}{\mu}\right)^k a_{i+1}(N-k,x,y,\mu) \\
&\quad+2y\sum_{k=0}^{N}\left(\frac{x\log(1+\mu)}{\mu}\right)^k a_{i-1}(N-k,x,y,\mu), (1\le i\le N-1).
\end{aligned} \tag{57}$$

Here is a plot of the surface for this solution. In the left picture of **Figure 1**, we choose $-2\le x\le 2,\ -1\le t\le 1, \mu=1/10,$ and $y=0.1$. In the right picture of **Figure 1**, we choose $-2\le y\le 2,\ -1\le t\le 1, \mu=1/10,$ and $x=0.1$.

Making N-times derivative for (10) with respect to t, we have

$$\left(\frac{\partial}{\partial t}\right)^N G(t,x,y,\mu) = \sum_{m=0}^{\infty}\mathrm{H}_{m+N}(x,y,\mu)\frac{t^m}{m!}. \tag{58}$$

By (58) and Theorem 5, we have

$$\begin{aligned}
&a_0(N,x,y,\mu)G(t,x,y,\mu) \\
&+a_1(N,x,y,\mu)tG(t,x,y,\mu) \\
&+\cdots \\
&+a_N(N,x,y,\mu)t^N G(t,x,y,\mu) \\
&=\sum_{m=0}^{\infty}\mathrm{H}_{m+N}(x,y,\mu)\frac{t^m}{m!}.
\end{aligned} \tag{59}$$

Hence we have the following theorem.

Theorem 6. For $N=0,1,2,\ldots,$ we get

$$\mathrm{H}_{m+N}(x,y,\mu) = \sum_{i=0}^{m}\frac{\mathrm{H}_{m-i}(x)a_i(N,x,y,\mu)m!}{(m-i)!}. \tag{60}$$

If we take $m=0$ in (60), then we have the below corollary.

Corollary 7. For $N=0,1,2,\ldots,$ we have

$$\mathrm{H}_N(x,y,\mu) = a_0(N,x,y,\mu), \tag{61}$$

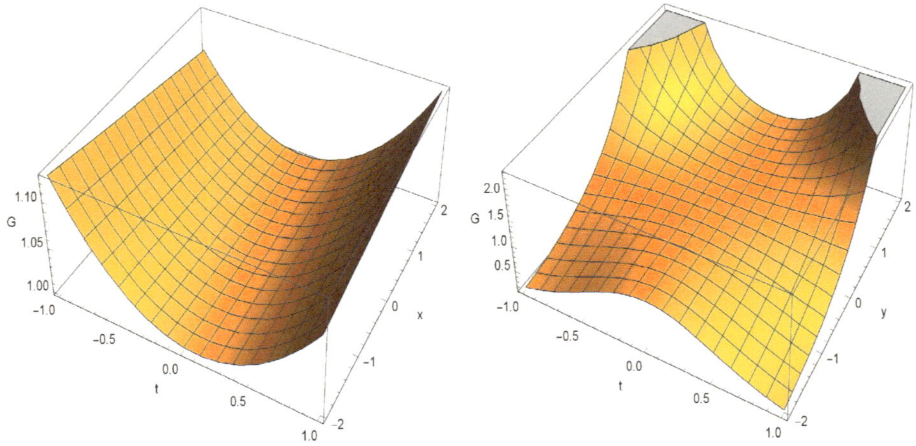

Figure 1.
The surface for the solution $G(t, x, y, \mu)$.

where

$$\begin{aligned} a_0(0, x, y, \mu) &= 1, \\ a_0(N+1, x, y, \mu) &= \sum_{i=0}^{N} \left(\frac{x \log(1+\mu)}{\mu} \right)^i a_1(N-i, x, y, \mu) \\ &\quad + \left(\frac{\log(1+\mu)}{\mu} \right)^{N+1} x^{N+1}. \end{aligned} \tag{62}$$

The first few of them are

$$\begin{aligned} \mathrm{H}_0(x, y, \mu) &= 1, \\ \mathrm{H}_1(x, y, \mu) &= \frac{\log(1+\mu)}{\mu} x, \\ \mathrm{H}_2(x, y, \mu) &= 2y + \frac{(\log(1+\mu))^2}{\mu^2} x^2, \\ \mathrm{H}_3(x, y, \mu) &= 6xy \frac{\log(1+\mu)}{\mu} + \frac{(\log(1+\mu))^3}{\mu^3} x^3, \\ \mathrm{H}_4(x, y, \mu) &= 12y^2 + 12x^2 y \frac{(\log(1+\mu))^2}{\mu^2} + \frac{(\log(1+\mu))^4}{\mu^4} x^4, \\ \mathrm{H}_5(x, y, \mu) &= 60xy^2 \frac{\log(1+\mu)}{\mu} + 20x^3 y \frac{(\log(1+\mu))^3}{\mu^3} + \frac{(\log(1+\mu))^5}{\mu^5} x^5. \end{aligned} \tag{63}$$

5. Zeros of the 2-variable modified degenerate Hermite polynomials

This section shows the benefits of supporting theoretical prediction through numerical experiments and finding new interesting pattern of the zeros of the 2-variable modified degenerate Hermite equations $\mathrm{H}_n(x, y|\mu) = 0$. By using computer, the 2-variable modified degenerate Hermite polynomials $\mathrm{H}_n(x, y|\mu)$ can be determined explicitly. We investigate the zeros of the 2-variable modified degenerate

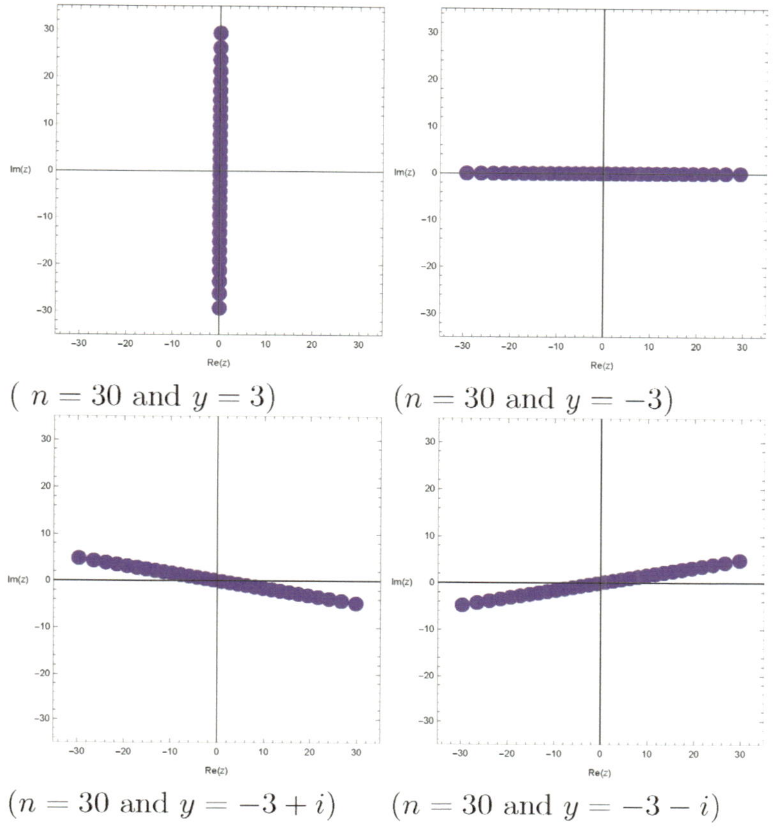

Figure 2.
Zeros of $\mathrm{H}_n(x, y|\mu) = 0$.

Hermite equations $\mathrm{H}_n(x, y|\mu) = 0$. The zeros of the $\mathrm{H}_n(x, y|\mu) = 0$ for $n = 30, y = 3, -3, 3 + i, -3 - i, \mu = 1/2$, and $x \in \mathbb{C}$ are displayed in **Figure 2**. In the top-left picture of **Figure 2**, we choose $n = 30$ and $y = 3$. In the top-right picture of **Figure 2**, we choose $n = 30$ and $y = -3$. In the bottom-left picture of **Figure 2**, we choose $n = 30$ and $y = -3 + i$. In the bottom-right picture of **Figure 2**, we choose $n = 30$ and $y = -3 - i$.

Stacks of zeros of the 2-variable modified degenerate Hermite equations $\mathrm{H}_n(x, y|\mu) = 0$ for $1 \le n \le 50, \mu = 1/2$ from a 3-D structure are presented in **Figure 3**. In the top-left picture of **Figure 3**, we choose $y = 3$. In the top-right picture of **Figure 3**, we choose $y = -3$. In the bottom-left picture of **Figure 3**, we choose $y = -3 + i$. In the bottom-right picture of **Figure 3**, we choose $y = -3 - i$.

Our numerical results for approximate solutions of real zeros of the 2-variable modified degenerate Hermite equations $\mathrm{H}_n(x, y|\mu) = 0$ are displayed (**Tables 1** and **2**).

We observed a remarkable regular structure of the complex roots of the 2-variable modified degenerate Hermite equations $\mathrm{H}_n(x, y|\mu) = 0$ and also hope to verify same kind of regular structure of the complex roots of the 2-variable modified degenerate Hermite equations $\mathrm{H}_n(x, y|\mu) = 0$ (**Table 1**).

Plot of real zeros of the 2-variable modified degenerate Hermite equations $\mathrm{H}_n(x, y|\mu) = 0$ for $1 \le n \le 50, \mu = 1/2$ structure are presented in **Figure 4**. In the top-left picture of **Figure 4**, we choose $y = 3$. In the top-right picture of **Figure 4**, we

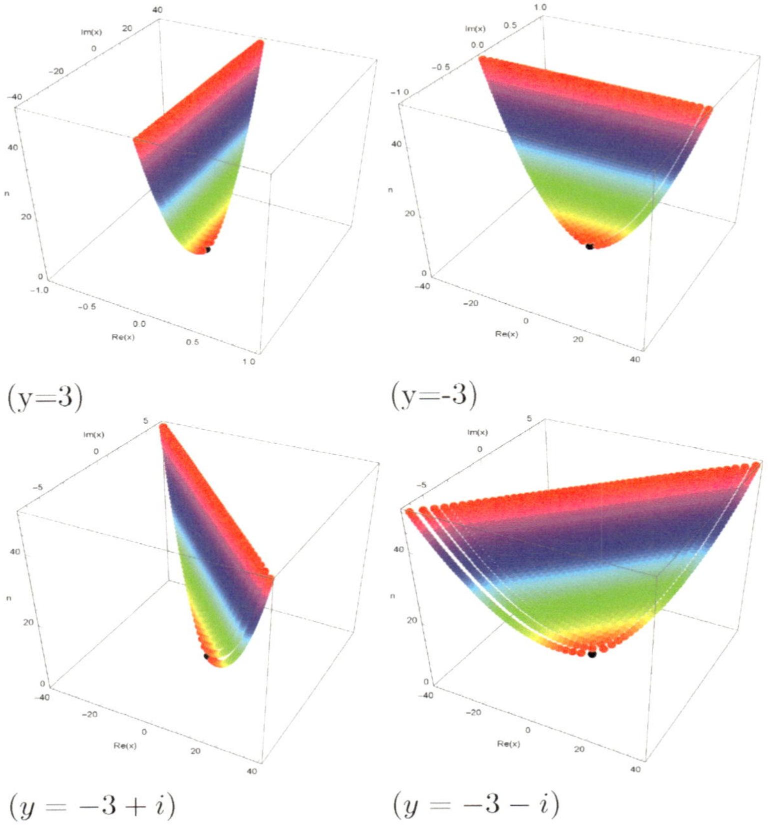

Figure 3.
Stacks of zeros of $H_n(x, y|\mu) = 0, 1 \le n \le 50$.

	$y = 3, \mu = 1/2$		$y = -3, \mu = 1/2$	
Degree n	**Real zeros**	**Complex zeros**	**Real zeros**	**Complex zeros**
1	1	0	1	0
2	0	2	2	0
3	1	2	3	0
4	0	4	4	0
5	1	4	5	0
6	0	6	6	0
7	1	6	7	0
8	0	8	8	0
9	1	8	9	0
10	0	10	10	0

Table 1.
Numbers of real and complex zeros of $H_n(x, y|\mu) = 0$.

Degree n	x
1	0
2	−3.0206, 3.0206
3	−5.2318, 0, 5.2318
4	−7.0513, − 2.2412, 2.2412, 7.0513
5	−8.6297, − 4.0948, 0, 4.0948, 8.6297
6	−10.041, − 5.7064, − 1.8628, 1.8628, 5.7064, 10.041
7	−11.329, − 7.1490, − 3.4870, 0, 3.4870, 7.1490, 11.329
8	−12.519, − 8.4652, − 4.9433, − 1.6283, 1.6283, 4.9433, 8.4652, 12.519

Table 2.
Approximate solutions of $\mathrm{H}_n(x, y|\mu) = 0, x \in \mathbb{R}$.

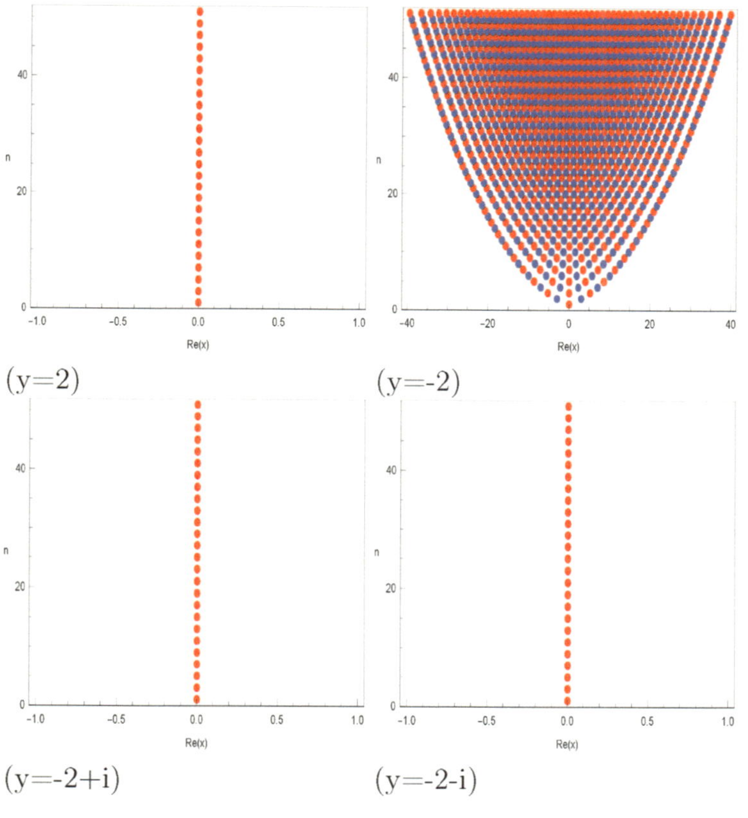

Figure 4.
Real zeros of $\mathrm{H}_n(x,y|\mu) = 0$ for $1 \le n \le 50$, $\mu = \frac{1}{2}$.

choose $y = -3$. In the bottom-left picture of **Figure 4**, we choose $y = -3 + i$. In the bottom-right picture of **Figure 4**, we choose $y = -3 - i$.

Next, we calculated an approximate solution satisfying $\mathrm{H}_n(x, y|\mu) = 0, x \in \mathbb{C}$. The results are given in **Table 2**. In **Table 2**, we choose $y = -3$ and $\mu = 1/2$.

6. Conclusions

In this chapter, we constructed the 2-variable modified degenerate Hermite polynomials and got some new symmetric identities for 2-variable modified degenerate Hermite polynomials. We constructed differential equations arising from the generating function of the 2-variable modified degenerate Hermite polynomials $\mathrm{H}_n(x,y|\mu)$. We also investigated the symmetry of the zeros of the 2-variable modified degenerate Hermite equations $\mathrm{H}_n(x,y|\mu)=0$ for various variables x and y. As a result, we found that the distribution of the zeros of 2-variable modified degenerate Hermite equations $\mathrm{H}_n(x,y|\mu)=0$ is very regular pattern. So, we make the following series of conjectures with numerical experiments:

Let us use the following notations. $R_{\mathrm{H}_n(x,y|\mu)}$ denotes the number of real zeros of $\mathrm{H}_n(x,y|\mu)=0$ lying on the real plane $Im(x)=0$ and $C_{\mathrm{H}_n(x,y|\mu)}$ denotes the number of complex zeros of $\mathrm{H}_n(x,y|\mu)=0$. Since n is the degree of the polynomial $\mathrm{H}_n(x,y|\mu)$, we have $R_{\mathrm{H}_n(x,y|\mu)}=n-C_{\mathrm{H}_n(x,y|\mu)}$.

We can see a good regular pattern of the complex roots of the 2-variable modified degenerate Hermite equations $\mathrm{H}_n(x,y,\mu)=0$ for y and μ. Therefore, the following conjecture is possible.

Conjecture 1. Let n be odd positive integer. For $a>0$ or $a\in\mathbb{C}\backslash\{a\mid a<0\}$, prove or disprove that

$$R_{\mathrm{H}_n(x,a,\mu)}=1,\quad C_{\mathrm{H}_n(x,a,\mu)}=2\left[\frac{n}{2}\right], \tag{64}$$

where $\mathbb{C}$ is the set of complex numbers.

Conjecture 2. Let n be odd positive integer and $a\in\mathbb{C}$. Prove or disprove that

$$\mathrm{H}_n(0,a,\mu)=0. \tag{65}$$

As a result of investigating more y and μ variables, it is still unknown whether the conjecture 1 and conjecture 2 is true or false for all variables y and μ.

We observe that solutions of the 2-variable modified degenerate Hermite equations $\mathrm{H}_n(x,y,\mu)=0$ has not $Re(x)=b$ reflection symmetry for $b\in\mathbb{R}$. It is expected that solutions of the 2-variable modified degenerate Hermite equations $\mathrm{H}_n(x,y,\mu)=0$, has not $Re(x)=b$ reflection symmetry (see **Figures 2–4**).

Conjecture 3. Prove that the zeros of $\mathrm{H}_n(x,a,\mu)=0, a\in\mathbb{R}$, has $Im(x)=0$ reflection symmetry analytic complex functions. Prove that the zeros of $\mathrm{H}_n(x,a,\mu)=0, a<0, a\in\mathbb{C}\backslash\mathbb{R}$, has not $Im(x)=0$ reflection symmetry analytic complex functions.

Finally, we consider the more general problems. How many zeros does $\mathrm{H}_n(x,y,\mu)$ have? We are not able to decide if $\mathrm{H}_n(x,y,\mu)=0$ has n distinct solutions. We would like to know the number of complex zeros $C_{\mathrm{H}_n(x,y,\mu)}$ of $\mathrm{H}_n(x,y,\mu)=0$.

Conjecture 4. For $a\in\mathbb{C}$, prove or disprove that $\mathrm{H}_n(x,a,\mu)=0$ has n distinct solutions.

As a result of investigating more n variables, it is still unknown whether the conjecture is true or false for all variables n (see **Tables 1** and **2**).

We expect that research in these directions will make a new approach using the numerical method related to the research of the 2-variable modified degenerate Hermite equations $\mathrm{H}_n(x,y,\mu)=0$ which appear in applied mathematics and mathematical physics.

Additional information

Mathematics Subject Classification: 05A19, 11B83, 34A30, 65L99

Author details

Cheon Seoung Ryoo
Department of Mathematics, Hannam University, Daejeon, Republic of Korea

*Address all correspondence to: ryoocs@hnu.kr

References

[1] Andrews LC. Special Functions for Engineers and Mathematicians. New York, NY, USA: Macmillan. Co.; 1985

[2] Appell P, Hermitt Kampéde Fériet J. Fonctions Hypergéométriques et Hypersphériques: Polynomes d Hermite. Paris, France: Gauthier-Villars; 1926

[3] Erdelyi A, Magnus W, Oberhettinger F, Tricomi FG. Higher Transcendental Functions. Vol. 3. New York, NY, USA: Krieger; 1981

[4] Andrews GE, Askey R, Roy R. Special Functions. Cambridge, England: Cambridge University Press; 1999

[5] Arfken G. Mathematical Methods for Physicists. 3rd ed. Orlando, FL: Academic Press; 1985

[6] Carlitz L. Degenerate Stirling, Bernoulli and Eulerian numbers. Utilitas Mathematica. 1979;**15**:51-88

[7] Young PT. Degenerate Bernoulli polynomials, generalized factorial sums, and their applications. Journal of Number Theory. 2008;**128**:738-758

[8] Ryoo CS. A numerical investigation on the structure of the zeros of the degenerate Euler-tangent mixed-type polynomials. Journal of Nonlinear Sciences and Applications. 2017;**10**: 4474-4484

[9] Ryoo CS. Notes on degenerate tangent polynomials. The Global Journal of Pure and Applied Mathematics. 2015; **11**:3631-3637

[10] Hwang KW, Ryoo CS. Differential equations associated with two variable degenerate Hermite polynomials. Mathematics. 2020;**8**:1-18. DOI: 10.3390/math8020228

[11] Hwang KW, Ryoo CS, Jung NS. Differential equations arising from the generating function of the (r,β)-Bell polynomials and distribution of zeros of equations. Mathematics. 2019;**7**:1-11. DOI: 10.3390/math7080736

[12] Ryoo CS. Differential equations associated with tangent numbers. Journal of Applied Mathematics and Informatics. 2016;**34**:487-494

[13] Ryoo CS. Some identities involving Hermitt Kampé de Fériet polynomials arising from differential equations and location of their zeros. Mathematics. 2019;7:1-11. DOI: 10.3390/math7010023

[14] Ryoo CS, Agarwal RP, Kang JY. Differential equations associated with Bell-Carlitz polynomials and their zeros. Neural, Parallel & Scientific Computations. 2016;**24**:453-462

Chapter 11

Elliptic Curve over a Local Finite Ring R_n

Abdelhakim Chillali and Lhoussain El Fadil

Abstract

The goal of this chapter is to study some arithmetic proprieties of an elliptic curve defined by a Weierstrass equation on the local ring $R_n = \mathbb{F}_q[X]/(X^n)$, where $n \geq 1$ is an integer. It consists of, an introduction, four sections, and a conclusion. In the first section, we review some fundamental arithmetic proprieties of finite local rings R_n, which will be used in the remainder of the chapter. The second section is devoted to a study the above mentioned elliptic curve on these finite local rings for arbitrary characteristics. A restriction to some specific characteristic cases will then be considered in the third section. Using these studies, we give in the fourth section some cryptography applications, and we give in the conclusion some current research perspectives concerning the use of this kind of curves in cryptography. We can see in the conclusion of research in perspectives on these types of curves.

Keywords: elliptic curve, finite ring, cryptography

1. Introduction

Elliptic curves are especially important in number theory and constitute a major area of current research; for example, they were used in Andrew Wiles's proof of Fermat's Last Theorem. They also find applications in elliptic curve cryptography (ECC), integer factorization, classical mechanics in the description of the movement of spinning tops, to produce efficient codes ... For these reasons, the subject is well known, presented, and worth exploring.

The purpose of cryptography is to ensure the security of communications and data stored in the presence of adversaries [1–3]. It offers a set of techniques for providing confidentiality, authenticity, and integrity services. Cryptology, also known as the science of secrecy, combines cryptography and cryptanalysis. While the role of cryptographers is to design, build, and prove cryptosystems, among other things, the goal of cryptanalysis is to "break" these systems. The history of cryptography has long been the history of secret codes and along all previous times, this has affected the fate of men and nations [4]. In fact, until 1970, the main goal of cryptography was to build a signature encryption systems [5, 6], but thanks to cryptanalysis, the army and the black cabinets of diplomats were able to wage their wars in the shadows controlling the communication networks, especially of their enemies [7, 8]. The internet revolution and the increasingly massive use of information in digital form facilitated communications but in counterparty it weakened the security level of information. Indeed, "open" networks create security holes, which allow access to the information. Cryptography, or the art of encrypting messages, a science that sites today in the crossroads of mathematics, computer sciences, and some applied physics, has then become a necessity for today's civilization to keep its secrets from adversaries. Confusion is often made between cryptography and cryptology, but the difference exists. Cryptology is the "science of secrecy," and combines two branches on the one hand, cryptography, which makes it possible to encrypt messages, and on the other hand, cryptanalysis, which serves to decrypt them. Our focus in this chapter is to show how some elliptic curves, mathematical objects studied particularly in algebraic geometry [9–12]. You can give several definitions depending on the person you are talking to. Cryptography indeed used elliptic curves for more than 40 years the appearance of the Diffie-Hellman key exchange protocol and the ElGamal cryptogram [13–15]. These cryptographic protocols use in particular group structures, for by applying these methods to groups defined by elliptic curves, a new speciality was born at the end of the 1980: ECC, Elliptic Curve Cryptography. Recall that Diffie-Hellman key exchange which is based on the difficulty of the discrete logarithm problem (DLP) [16–18]. The success of elliptic curves in public key cryptographic systems has then created a new interest in the study of the arithmetic of these geometric objects. The group of points on an elliptical curve is an interesting group in cryptography because there is no known sub-exponential algorithm for sound (DLP) [19–21]. In general, the DLP is difficult to be solved, but not as much as in a generic group as in the case of finite field. We know sub-exponential algorithms to solve it depending on the size of the group to use, which impose criteria for the PLD to be infeasible. The prime number p which is the characteristic of our base ring must then have at least 1024 bits, which offers a security level similar to the one given by a generic order group of 160 bits. Recall that a generic group for the DLP is a group for which there is no a specific algorithm to solve the DLP [22], so that the only available algorithms are those for all groups.

In [23], Elhassani et al. have built an encryption method based on DLP and Lattice. Boulbot et al. in [24] have studied elliptic curves on a non-local ring to compare these curves on local and non-local rings, while in [25], Sahmoudi et al. have studied these types of curves on a family of finite rings in the authors have introduced a cryptosystem on these types of curves, see [26].

In this chapter, d and n are a positive integers and $q = p^d$ is a power of a prime natural number p.

2. The ring $R_n = \mathbb{F}_q[X]/(X^n)$

Let $R_n = \mathbb{F}_q[X]/(X^n)$ be a $\mathbb{F}_q$-algebra of dimension n, with $(1, \epsilon, \ldots, \epsilon^{n-1})$ as a $\mathbb{F}_q$-basis, where $\epsilon = \overline{X}$, $\epsilon^n = 0$, $\mathbb{F}_q$ is the finite field of order $q = p^r$, and p being a prime integer [27–29].

2.1 Internal laws in R_n

Recall that the two laws "+" and "." are naturally defined on R_n [30, 31]: for every two elements $X = \sum_{i=0}^{n-1} x_i \epsilon^i$ and $Y = \sum_{i=0}^{n-1} y_i \epsilon^i$ in R_n, with $x_1, \ldots, x_n, y_1, \ldots, y_n$ in $\mathbb{F}_q$,

$$X + Y = \sum_{i=0}^{n-1} z_i \epsilon^i, \text{ where } z_j = x_j + y_j \text{ in } \mathbb{F}_q \quad (1)$$

DOI: http://dx.doi.org/10.5772/intechopen.93476

$$X.Y = \sum_{i=0}^{n-1} z_i \epsilon^i, \text{ where } z_j = \sum_{i=0}^{j} x_i y_{j-i} \text{(The cauchy product)} \tag{2}$$

Corollary 2.1 *Let* $X = \sum_{i=0}^{n-1} x_i \epsilon^i \in R_n$, *then* $X^2 = \sum_{i=0}^{n-1} x'_i \epsilon^i$ *where*

$$\forall k \geq 0, \begin{cases} x'_{2k} = x_k^2 + 2\sum_{i=0}^{k-1} x_i x_{2k-i} \\ x'_{2k+1} = 2\sum_{i=0}^{k} x_i x_{2k+1-i} \end{cases} \tag{3}$$

Proof.
By formula (2), we have

$$\forall j \geq 0, x'_j = \sum_{i=0}^{j} x_i x_{j-i}. \tag{4}$$

$$\text{For } j = 2k, x'_{2k} = \sum_{i=0}^{2k} x_i x_{2k-i}, \tag{5}$$

$$\text{so, } x'_{2k} = x_k^2 + 2\sum_{i=0}^{k-1} x_i x_{2k-i}. \tag{6}$$

$$\text{Similarly, for } j = 2k+1, x'_{2k+1} = \sum_{i=0}^{2k+1} x_i x_{2k+1-i}, \tag{7}$$

$$\text{then, } x'_{2k+1} = 2\sum_{i=0}^{k} x_i x_{2k+1-i}. \tag{8}$$

Under the same hypotheses of the corollary (2.1) and by an analogous proof, we have the following corollary:

Corollary 2.2 $X^3 = \sum_{i=0}^{n-1} x''_i \epsilon^i$, *where*

$$\forall k \geq 0, \begin{cases} x''_{2k} = x'_{2k} x_0 + \sum_{l=0}^{k-1} \left(x'_{2l} x_{2k-2l} + x'_{2l+1} x_{2k-1-2l} \right) \\ x''_{2k+1} = \sum_{l=0}^{k} \left(x'_{2l} x_{2k+1-2l} + x'_{2l+1} x_{2k-2l} \right) \end{cases} \tag{9}$$

Lemma 2.3 *Let* $Y = \sum_{i=0}^{n-1} y_i \epsilon^i$ *the inverse of* $X = \sum_{i=0}^{n-1} x_i \epsilon^i$. *Then*

$$\begin{cases} y_0 = x_0^{-1} \\ y_j = -x_0^{-1} \sum_{i=0}^{j-1} y_i x_{j-i}, \quad \forall j > 0 \end{cases} \tag{10}$$

Proof.

Let $Y = \sum_{i=0}^{n-1} y_i \epsilon^i$ be the inverse of $X = \sum_{i=0}^{n-1} x_i \epsilon^i$. Then $XY = 1$, by formula (2), we have

$$XY = \sum_{i=0}^{n-1} z_i \epsilon^i, \text{ where } z_j = \sum_{i=0}^{j} x_i y_{j-i}. \tag{11}$$

So,

$$z_0 = 1 \text{ and } \forall j > 0, z_j = 0, \tag{12}$$

which means that,

$$\begin{cases} y_0 = x_0^{-1} \\ y_j = -x_0^{-1}\sum_{i=0}^{j-1} y_i x_{j-i}, \qquad \forall j > 0 \end{cases} \tag{13}$$

Lemma 2.4 *The non inverse elements in R_n are the elements of the form $\sum_{i=1}^{n-1} x_i \epsilon^i$ where $x_i \in \mathbb{F}_q^{n-1}$ for all $1 \leq i \leq n-1$.*

Proof.

Let $X = \sum_{i=0}^{n-1} x_i \epsilon^i \in R_n$. By lemma (2.3), X is invertible in R_n if and only if x_0 is invertible in $\mathbb{F}_q$. As $\mathbb{F}_q$ is a field, this means $x_0 \neq 0$.

Corollary 2.5 *The ring R_n is local, with maximal ideal $I_n = \epsilon R_n$.*

Notation.

Let $k \geq 2$, we denote:

1.

$$\pi_k : \left| \begin{array}{ccc} R_k & \rightarrow & R_{k-1} \\ \sum_{i=0}^{k-1} x_i \epsilon^i & \mapsto & \sum_{i=0}^{k-2} x_i \delta^i \end{array} \right.$$

the projection of R_k on R_{k-1}.

2.

$$k^{\pi} : \left| \begin{array}{ccc} R_k & \rightarrow & R_1 \\ \sum_{i=0}^{k-1} x_i \epsilon^i & \mapsto & x_0 \end{array} \right.$$

the canonical projection of R_k on $R_1 = \mathbb{F}_q$.

Corollary 2.6 *π_k et k^{π} are two ring homomorphisms.*

Proof.

We have,

$$\begin{aligned} \pi_k\left(\sum_{i=0}^{k-1} x_i \epsilon^i + \sum_{i=0}^{k-1} y_i \epsilon^i\right) &= \pi_k\left(\sum_{i=0}^{k-1} (x_i + y_i)\epsilon^i\right) \\ &= \sum_{i=0}^{k-2} (x_i + y_i)\delta^i \\ &= \pi_k\left(\sum_{i=0}^{k-1} x_i \epsilon^i\right) + \pi_k\left(\sum_{i=0}^{k-1} y_i \epsilon^i\right) \end{aligned} \tag{14}$$

and

$$\left(\sum_{i=0}^{k-1} x_i \epsilon^i\right)\left(\sum_{i=0}^{k-1} y_i \epsilon^i\right) = \sum_{i=0}^{k-1} z_i \epsilon^i, \text{where } z_j = \sum_{i=0}^{j} x_i y_{j-i}.$$

$$\pi_k\left(\sum_{i=0}^{k-1} z_i \epsilon^i\right) = \sum_{i=0}^{k-2} z_i \delta^i \tag{15}$$

$$\pi_k\left(\sum_{i=0}^{k-1} x_i \epsilon^i\right)\pi_k\left(\sum_{i=0}^{k-1} y_i \epsilon^i\right) = \sum_{i=0}^{k-2} z_i \delta^i$$

Note that in addition, for every $k \geq 1$,

$$k^{\pi} = \pi_2 \circ \pi_3 \circ \pi_4 \ldots\ldots \circ \pi_k \tag{16}$$

So, π_k and k^{π} are tow rings morphisms.

Theorem 2.7 *Let* $n \geq 2$ *be an integer,* $a = \tilde{a} + a_{n-1}\epsilon^{n-1}$, $b = \tilde{b} + b_{n-1}\epsilon^{n-1}$, $X = \tilde{X} + x_{n-1}\epsilon^{n-1}$, $Y = \tilde{Y} + y_{n-1}\epsilon^{n-1}$ *and* $Z = \tilde{Z} + z_{n-1}\epsilon^{n-1}$ *be elements of* R_n *with:*

$$Y^2Z = X^3 + aXZ^2 + bZ^3. \tag{17}$$

Then

$$\tilde{Y}^2\tilde{Z} = \tilde{X}^3 + \tilde{a}\tilde{X}\tilde{Z}^2 + \tilde{b}\tilde{Z}^3 + \left[D - \left(Ay_{n-1} + Bz_{n-1} + Cx_{n-1}\right)\right]\epsilon^{n-1} \tag{18}$$

where,

$$A = 2y_0z_0, \tag{19}$$

$$B = y_0^2 - 3z_0^2b_0 - 2z_0a_0x_0, \tag{20}$$

$$C = -\left(3x_0^2 + a_0z_0^2\right) \tag{21}$$

and

$$D = b_{n-1}z_0^3 + a_{n-1}x_0z_0^2. \tag{22}$$

Proof.
We have:

$$\begin{aligned}
Y^2Z &= \left(\tilde{Y} + y_{n-1}\epsilon^{n-1}\right)^2\left(\tilde{Z} + z_{n-1}\epsilon^{n-1}\right) \\
&= \tilde{Y}^2\tilde{Z} + \left(y_0^2z_{n-1} + 2y_0z_0y_{n-1}\right)\epsilon^{n-1} \\
X^3 &= \left(\tilde{X} + x_{n-1}\epsilon^{n-1}\right)^3 \\
&= \tilde{X}^3 + 3x_0^2x_{n-1}\epsilon^{n-1} \\
aXZ^2 &= \tilde{a}\tilde{X}\tilde{Z}^2 + \left(2z_{n-1}z_0a_0x_0 + a_0x_{n-1}z_0^2 + a_{n-1}x_0z_0^2\right)\epsilon^{n-1} \\
bZ^3 &= \tilde{b}\tilde{Z}^3 + \left(b_{n-1}z_0^3 + 3z_0^2z_{n-1}b_0\right)\epsilon^{n-1}
\end{aligned} \tag{23}$$

If

$$Y^2Z = X^3 + aXZ^2 + bZ^3, \tag{24}$$

then, $\tilde{Y}^2\tilde{Z} = \tilde{X}^3 + \tilde{a}\tilde{X}\tilde{Z}^2 + \tilde{b}\tilde{Z}^3 + (3x_0^2x_{n-1} + 2z_{n-1}z_0a_0x_0 + a_0x_{n-1}z_0^2 + a_{n-1}x_0z_0^2 3z_0^2z_{n-1}b_0 - y_0^2z_{n-1} - 2y_0z_0y_{n-1})\epsilon^{n-1}$ and therefore,

$$\tilde{Y}^2\tilde{Z} = \tilde{X}^3 + \tilde{a}\tilde{X}\tilde{Z}^2 + \tilde{b}\tilde{Z}^3 + \left[D - \left(Ay_{n-1} + Bz_{n-1} + Cx_{n-1}\right)\right]\epsilon^{n-1} \tag{25}$$

where,

$$A = 2y_0z_0, \tag{26}$$

$$B = y_0^2 - 3z_0^2b_0 - 2z_0a_0x_0, \tag{27}$$

$$C = -\left(3x_0^2 + a_0z_0^2\right) \tag{28}$$

and

$$D = b_{n-1}z_0^3 + a_{n-1}x_0z_0^2. \tag{29}$$

2.2 Primitive triples

Definition 2.8 *Let R be a ring. We say that an element* $(x, y, z) \in R^3$ *is primitive if:* $xR + yR + zR = R$. *The set of these primitive triplets will be denoted* $\mathcal{P}(R)$.

Remark 2.9 *The equality* $xR + yR + zR = R$ *means that there exists* $(\alpha, \beta, \lambda) \in R^3$ *such that* $1_R = \alpha x + \beta y + \lambda z$.

Proposition 2.10 *Let R be a local ring, then* $(x, y, z) \in R^3$ *is a primitive triple if and only if at least one of the elements x, y, and z is invertible in R.*

Proof.

Suppose that x, y and z are not invertible in R, then:
$(x, y, z) \in \mathcal{M}^3$ where $\mathcal{M}$ is the unique maximal ideal of R, hence

$$xR + yR + zR \subset \mathcal{M} \subsetneq R, \tag{30}$$

which contradicts that (x, y, z) is a primitive triple.

Conversely, suppose, for example, that x is invertible in R, then $xR = R$, so $xR + yR + zR = R$.

Remark 2.11 *If R is a field, then an element* $(x, y, z) \in R^3$ *is primitive if and only if* $(x, y, z) \neq (0, 0, 0)$.

2.3 The projective plane on a finite ring

Let R is a ring. The projective plane on R is the set of equivalence classes of $\mathcal{P}(R)$ modulo; the equivalence relation $\sim R$ defined by:

$$\left(x_1, y_1, z_1\right) \sim R\left(x_2, y_2, z_2\right) \Leftrightarrow \exists \lambda \in R^\times : \ \left(x_2, y_2, z_2\right) = \lambda\left(x_1, y_1, z_1\right). \tag{31}$$

We denote the projective plane on R by $\mathbb{P}^2(R)$, it is the quotient set $\frac{\mathcal{P}(R)}{\sim R}$, and we write $[x : y : z]$ for the equivalence class of $(x, y, z) \in \mathcal{P}(R)$. Thus, we have:

$$\left[x_1 : y_1 : z_1\right] = \left[x_2 : y_2 : z_2\right] \Leftrightarrow \exists \lambda \in R^\times : x_2 = \lambda x_1, y_2 = \lambda y_1 \text{ and } z_2 = \lambda z_1. \tag{32}$$

Example 2.12 *We Consider the finite ring* $\mathbb{F}_2[e] = \{\alpha + \beta e / \alpha \in \mathbb{F}_2 \, and \, \beta \in \mathbb{F}_2\}$, *where e is an indeterminate satisfying* $e^2 = 0$. *The group of units for this ring is* $(\mathbb{F}_2[e])^\times = \{1, 1 + e\}$.

As this ring is local with maximal ideal $e\mathbb{F}_2[e]$, *then an element* (x, y, z) *of* $\mathbb{F}_2[e]^3$ *is non primitive if and only if* $(x, y, z) \in \{0, e\}^3$. *As one can see, there are eight elements which are not primitive, and therefore the set* $\mathcal{P}(\mathbb{F}_2[e])$ *contains* $64 - 8 = 56$ *primitive triples as given below:*

$$\begin{aligned}\mathcal{P}(\mathbb{F}_2[e]) = \{&(0,0,1),(0,1,1+e),(0,1,0),(0,1,1),(0,1,e),(0,1,1+e),(0,e,1),\\ &(0,e,1+e),(0,1+e,0),\ (0,1+e,1),\ (0,1+e,e),(0,1+e,1+e),\\ &(1,0,0),(1,0,1),(1,0,e),(1,0,1+e),(1,1,0),(1,1,1),(1,1,e),\\ &(1,1,1+e),(1,e,0),(1,e,1),(1,e,e),(1,e,1+e),(1,1+e,0),\\ &(1,1+e,1),(1,1+e,e),(1,1+e,1+e),(e,0,1),(e,0,1+e),(e,1,0),\\ &(e,1,1),(e,1,e),(e,1,1+e),(e,e,1),(e,e,1+e),(e,1+e,0),(e,1+e,1),\\ &(e,1+e,e),(e,1+e,1+e),(1+e,0,0),(1+e,0,1),(1+e,0,e),\\ &(1+e,0,1+e),(1+e,1,0),(1+e,1,1),(1+e,1,e),(1+e,1,1+e),\\ &(1+e,e,0),(1+e,e,1),(1+e,e,e),(1+e,e,1+e),(1+e,1+e,0),\\ &(1+e,1+e,1),(1+e,1+e,e),(1+e,1+e,1+e)\}.\end{aligned} \tag{33}$$

Let (x,y,z) and (x',y',z') be two elements in $\mathcal{P}(\mathbb{F}_2[e])$, then:
$[x':y':z'] = [x:y:z] \Leftrightarrow (x',y',z') = (x,y,z)$ *or* $(x',y',z') = (x+xe, y+ye, z+ze)$
so every class in $\mathbb{P}^2(\mathbb{F}_2[e])$ contains two representatives, that is, the projective plane $\mathbb{P}^2(\mathbb{F}_2[e])$ contains exactly the following 28 elements:

$$\begin{aligned}\mathbb{P}^2(\mathbb{F}_2[e]) =\{&[0:1:0],[0:0:1],[0:1:1],[0:1:e],[0:1:1+e],[0:e:1],[1:0:0],\\ &[1:0:1],[1:0:e],[1:0:1+e],[1:1:0],[1:1:1],[1:1:e],[1:1:1+e],\\ &[1:e:0],[1:e:1],[1:e:e],[1:e:1+e],[1:1+e:0],[1:1+e:1],\\ &[1:1+e:e],[1:1+e:1+e],[e:0:1],[e:1:0],[e:1:1],[e:1:e],\\ &[e:1:1+e],[e:e:1]\}.\end{aligned} \tag{34}$$

3. Elliptic curve over R_n

In this section, we study the elliptic curves defined on finite local rings R_n of characteristic a prime number p;

1. A projective Weierstrass equation on R_n is an equation of the form:

$$\mathbf{E}:\ Y^2Z + a_1XYZ + a_3YZ^2 = X^3 + a_2X^2Z + a_4X + a_6Z^3 \tag{35}$$

2. A affine Weierstrass equation on R_n is an equation of the form:

$$\mathbf{E}':\ Y^2 + a_1XY + a_3Y = X^3 + a_2X^2 + a_4X + a_6 \tag{36}$$

where $(a_1,a_2,a_3,a_4,a_6) \in {R_n}^5$.

3.1 Elliptic curve form

To an affine (or projective) Weierstrass Eqs. (3) and (4), we associate the following quantities:

$$\begin{aligned} b_2 &= a_1^2 + 4a_2 \\ b_4 &= 2a_4 + a_1a_3 \\ b_6 &= a_3^2 + 4a_6 \\ b_8 &= a_1^2a_6 + 4a_2a_6 - a_1a_3a_4 + a_2a_3^2 - a_4^2 \\ c_4 &= b_2^2 - 24b_4 \\ \Delta &= -b_2^2b_8 - 8b_4^3 - 27b_6^2 + 9b_2b_4b_6 \\ j &= \frac{c_4^3}{\Delta} \textit{ if } \Delta \neq 0 \end{aligned} \tag{37}$$

Δ is called the discriminant of **E** and j its j – invariant.

Remark 3.1 *On the field $R_1 = \mathbb{F}_q$, we denote the discriminant by Δ_0 and the j-invariant by j_0, while on the ring R_n, $n > 1$ we denote the discriminant by $\Delta_{\varepsilon,n}$ and the j-invariant by $j_{\varepsilon,n}$.*

We have $n^{\pi}(\Delta_{\varepsilon,n}) = \Delta_0$ and $n^{\pi}\left(j_{\varepsilon,n}\right) = j_0$.

Definition 3.2 *Let R be a finite ring and let $a = (a_1, a_2, a_3, a_4, a_6) \in R^5$. An elliptic curve on R corresponding to a, which we write $E_a(R)$, is the set of zeros in the projective plane $\mathbb{P}^2(R)$ of the Weierstrass Eq. (3), for which the discriminant Δ is invertible in R.*

Remark 3.3 *According to the characteristic of the ring R; $chra(R)$ we have the following cases:*

1. If $char(R) \neq 2$ and $char(R) \neq 3$, then:

$$E_{a,b}(R) = \left\{[X : Y : Z] \in \mathbb{P}^2(R)/Y^2Z = X^3 + aXZ^2 + bZ^3\right\} \tag{38}$$

for $(a,b) \in R \times R$, with $\Delta = \Delta_{a,b} = -16\left(4a^3 + 27b^2\right) \in R^{\times}$.

2. If $char(R) = 2$, then $E_{a,b}(R)$ has one of the following forms:

$$E_{a,b}(R) = \left\{[X : Y : Z] \in \mathbb{P}^2(R)/Y^2Z + XYZ = X^3 + aX^2Z + bZ^3\right\} \tag{39}$$

for $(a,b) \in R^2$, with $\Delta = \Delta_{a,b} = b \in R^{\times}$.
Or:

$$E_a(R) = \left\{[X : Y : Z] \in \mathbb{P}^2(R)/Y^2Z + a_1XYZ + a_3YZ^2 = X^3 + a_4XZ^2 + a_6Z^3\right\} \tag{40}$$

for $a = (a_1, a_3, a_4, a_6) \in R^4$, with a_1 non invertible and

$$\Delta = \Delta_a = a_1^3\left(a_1^3a_6 + a_1^2a_3a_4 + a_1a_4^2 + a_3^3\right) + a_3^4 \in R^{\times}. \tag{41}$$

3. If $char(R) = 3$, then $E_{a,b}(R)$ has one of the following forms:

$$E_{a,b}(R) = \left\{[X : Y : Z] \in \mathbb{P}^2(R)/Y^2Z = X^3 + aX^2Z + bZ^3\right\} \tag{42}$$

for $(a,b) \in R^2$, with $\Delta = \Delta_{a,b} = -a^3b \in R^{\times}$.
Or:

$$E_{a,b}(R) = \left\{[X : Y : Z] \in \mathbb{P}^2(R)/Y^2Z = X^3 + aXZ^2 + bZ^3\right\} \tag{43}$$

for $(a,b) \in R^2$, with $\Delta = \Delta_{a,b} = -a^3 \in R^{\times}$.

Remark 3.4 *A projective elliptic curve on a field K has one of the following normal forms (**Table 1**):*

		Normal form	
$char(\mathbf{K}) \neq 2, 3$		$Y^2Z = X^3 + a_4XZ^2 + a_6Z^3$	
		$\Delta = -16(4a_4^3 + 27a_6^2)$	$j = 1728\frac{4a_4^3}{4a_4^3+27a_6^2}$
$char(\mathbf{K}) = 3$	$j = 0$	$Y^2Z = X^3 + a_4XZ^2 + a_6Z^3$	
		$\Delta = -a_4^3$	
	$j \neq 0$	$Y^2Z = X^3 + a_2X^2Z + a_6Z^3$	
		$\Delta = -a_2^3a_6$	$j = -\frac{a_2^3}{a_6}$
$char(\mathbf{K}) = 2$	$j = 0$	$Y^2Z + a_3YZ^2 = X^3 + a_4XZ^2 + a_6Z^3$	
		$\Delta = a_3^4$	
	$j \neq 0$	$Y^2Z + XYZ = X^3 + a_2X^2Z + a_6Z^3$	
		$\Delta = a_6$	$j = \frac{1}{a_6}$

Table 1.
Elliptic curve form on a field.

3.2 Projective coordinates and group law

In this subsection, we give in projective coordinates the formulas for adding the points on an elliptic curve defined by Eq. (3) on the ring R_n, according to the normal form.

Using Bosma and Lenstra's theorem see [32], we can deduce the explicit formulas for the commutative additive law of the group $E_a(R_n)$. The results are given in the next theorems following the values of the characteristic of ring R_n [33–36]. Let

$$[X_1 : Y_1 : Z_1] + [X_2 : Y_2 : Z_2] = [X_3 : Y_3 : Z_3]. \tag{44}$$

Theorem 3.5 *[Characteristic two case]:*

- *If* $[n^\pi(X_1) : n^\pi(Y_1) : n^\pi(Z_1)] = [n^\pi(X_2) : n^\pi(Y_2) : n^\pi(Z_2)]$, then:

$$\begin{aligned} X_3 = \; & X_1Y_1Y_2^2 + X_2Y_1^2Y_2 + X_2^2Y_1^2 + X_1X_2^2Y_1 + aX_1^2X_2Y_2 + aX_1X_2^2Y_1 + aX_1^2X_2^2 + \\ & bX_1Y_1Z_2^2 + bX_2Y_2Z_1^2 + bX_1^2Z_2^2 + bY_1Z_2^2Z_1 + bY_2Z_1^2Z_2 + bX_1Z_2^2Z_1. \end{aligned} \tag{45}$$

$$\begin{aligned} Y_3 = \; & Y_1^2Y_2^2 + X_2Y_1^2Y_2 + aX_1X_2^2Y_1 + a^2X_1^2X_2^2 + bX_1^2X_2Z_2 + bX_1X_2^2Z_1 + \\ & bX_1Y_1Z_2^2 + bX_1^2Z_2^2 + abX_2^2Z_1^2 + abX_1^2Z_2^2 + bY_1Z_1Z_2^2 + bX_1Z_1Z_2^2 + abX_1Z_1Z_2^2 + \\ & abX_2Z_1^2Z_2 + b^2Z_1^2Z_2^2 \end{aligned} \tag{46}$$

$$\begin{aligned} Z_3 = \; & X_1^2X_2Y_2 + X_1X_2^2Y_1 + Y_1^2Y_2Z_2 + Y_1Y_2^2Z_1 + X_1^2X_2^2 + Y_1^2X_2Z_2 + X_1^2Y_2Z_2 + \\ & aX_1^2Y_2Z_2 + aX_2^2Y_1Z_1 + X_1^2X_2Z_2 + aX_1X_2^2Z_1 + bY_1Z_1Z_2^2 + bY_2Z_1^2Z_2 + bX_1Z_1Z_2^2. \end{aligned} \tag{47}$$

- *If* $[n^\pi(X_1) : n^\pi(Y_1) : n^\pi(Z_1)] \neq [n^\pi(X_2) : n^\pi(Y_2) : n^\pi(Z_2)]$, *then:*

$$\begin{aligned} X_3 = \; & X_1Y_2^2Z_1 + X_2Y_1^2Z_2 + X_1^2Y_2Z_2 + X_2^2Y_1Z_1 + aX_1^2X_2Z_2 + aX_1X_2^2Z_1 + \\ & bX_1Z_1Z_2^2 + bX_2Z_1^2Z_2. \end{aligned} \tag{48}$$

$$Y_3 = X_1^2X_2Y_2 + X_1X_2^2Y_1 + Y_1^2Y_2Z_2 + Y_1Y_2^2Z_1 + X_1^2Y_2Z_2 + X_2^2Y_1Z_1 + aX_1^2Y_2Z_2 + aX_2^2Y_1Z_1 + aX_1^2X_2Z_2 + aX_1X_2^2Z_1 + bY_1Z_1Z_2^2 + bY_2Z_1^2Z_2 + bX_1Z_1Z_2^2 + bX_2Z_1^2Z_2. \quad (49)$$

$$Z_3 = X_1^2X_2Z_2 + X_1X_2^2Z_1 + Y_1^2Z_2^2 + Y_2^2Z_1^2 + X_1Y_1Z_2^2 + X_2Y_2Z_1^2 + aX_1^2Z_2^2 + aX_2^2Z_1^2. \quad (50)$$

Theorem 3.6 *[Characteristic three case]:*

- *If* $[n^\pi(X_1) : n^\pi(Y_1) : n^\pi(Z_1)] = [n^\pi(X_2) : n^\pi(Y_2) : n^\pi(Z_2)]$, *then:*

$$X_3 = Y_1Y_2^2X_1 + Y_1{}^2Y_2X_2 + 2aX_1{}^2X_2Y_2 + 2aX_1X_2{}^2Y_1 + 2Z_1Z_2{}^2abY_1 + 2Z_1{}^2Z_2abY_2. \quad (51)$$

$$Y_3 = Y_1{}^2Y_2{}^2 + 2a^2X_1{}^2X_2{}^2 + a^2bX_1Z_1Z_2{}^2 + a^2bX_2Z_1{}^2Z_2. \quad (52)$$

$$Z_3 = aX_1X_2(Y_1Z_2 + Y_2Z_1) + a(X_1Y_2 + X_2Y_1)(X_1Z_2 + X_2Z_1) + Y_1Y_2(Y_1Z_2 + Y_2Z_1). \quad (53)$$

- *If* $[n^\pi(X_1) : n^\pi(Y_1) : n^\pi(Z_1)] \neq [n^\pi(X_2) : n^\pi(Y_2) : n^\pi(Z_2)]$, *then:*

$$X_3 = 2X_1Y_2Y_1Z_2 + X_1Y_2{}^2Z_1 + 2X_2Y_1{}^2Z_2 + X_2Y_1Y_2Z_1 + 2aX_1{}^2X_2Z_2 + aX_1X_2{}^2Z_1. \quad (54)$$

$$Y_3 = 2Y_1{}^2Y_2Z_2 + Y_1Y_2{}^2Z_1 + 2aX_1X_2Y_1Z_2 + aX_1X_2Y_2Z_1 + 2aX_1{}^2Y_2Z_2 + aX_2{}^2Y_1Z_1. \quad (55)$$

$$Z_3 = 2Y_1{}^2Z_2{}^2 + Y_2{}^2Z_1{}^2 + aX_1{}^2Z_2{}^2 + 2aX_2{}^2Z_1{}^2. \quad (56)$$

Theorem 3.7 *[The case where the characteristic is different from two and from three]:*

- *If* $[n^\pi(X_1) : n^\pi(Y_1) : n^\pi(Z_1)] = [n^\pi(X_2) : n^\pi(Y_2) : n^\pi(Z_2)]$, *then:*

$$X_3 = Y_1^2X_2Z_2 - Z_1X_1Y_2^2a(Z_1X_2 + X_1Z_2)(Z_1X_2 - X_1Z_2) + (2Y_1Y_2 - 3bZ_1Z_2)(Z_1X_2 - X_1Z_2) \quad (57)$$

$$Y_3 = Y_1Y_2(Z_2Y_1 - Z_1Y_2) - a\left(X_1Y_1Z_2^2 - Z_1^2X_2Y_2\right) + (-2aZ_1Z_2 - 3X_1X_2)(X_2Y_1 - X_1Y_2) - 3bZ_1Z_2(Z_2Y_1 - Z_1Y_2) \quad (58)$$

$$Z_3 = (Z_1Y_2 + Z_2Y_1)(Z_2Y_1 - Z_1Y_2) + (3X_1X_2 + aZ_1Z_2)(Z_1X_2 - X_1Z_2) \quad (59)$$

- *If* $[n^\pi(X_1) : n^\pi(Y_1) : n^\pi(Z_1)] \neq [n^\pi(X_2) : n^\pi(Y_2) : n^\pi(Z_2)]$, *then:*

$$X_3 = (Y_1Y_2 - 6bZ_1Z_2)(X_2Y_1 + X_1Y_2) + \left(a^2Z_1Z_2 - 2aX_1X_2\right)(Z_1Y_2 + Z_2Y_1) - 3b\left(X_1Y_1Z_2^2 + Z_1^2X_2Y_2\right) - a\left(Y_1Z_1X_2^2 + X_1^2Y_2Z_2\right) \quad (60)$$

$$Y_3 = Y_1^2Y_2^2 + 3aX_1^2X_2^2 + \left(-a^3 - 9b^2\right)Z_1^2Z_2^2 - a^2(Z_1X_2 + X_1Z_2)^2 - 2a^2Z_1X_1Z_2X_2 + (9bX_1X_2 - 3abZ_1Z_2)(Z_1X_2 + X_1Z_2) \quad (61)$$

$$Z_3 = (Y_1Y_2 + 3bZ_1Z_2)(Z_1Y_2 + Z_2Y_1) + (3X_1X_2 + 2aZ_1Z_2)(X_2Y_1 + X_1Y_2) + a\left(X_1Y_1Z_2^2 + Z^1X_2Y_2\right). \quad (62)$$

4. Elliptic curve on R_n where $char(\mathbf{R_n}) \neq 2, 3$

The objective of this chapter is to study elliptic curves defined by a Weierstrass equation with coefficients in a ring R_n such that $char(\mathbf{R_n}) \neq 2, 3$. We denote it by $E_{a,b}^n$. Let

$$k_\theta : \left| \begin{array}{ccc} \mathbb{F}_q^{k-1} & \longrightarrow & E_{a,b}^k \\ (x_1, x_2, \ldots, x_{k-1}) & \mapsto & \left[\sum_{i=1}^{k-1} x_i \epsilon^i : 1 : \sum_{i=3}^{k-1} z_i \epsilon^i\right] \end{array} \right. \tag{63}$$

we denote, $k_G = k_\theta\left(\mathbb{F}_q^{k-1}\right)$.

4.1 The morphisms π^k et θ_k

Lemma 4.1 *The application*

$$\pi^k : \left| \begin{array}{c} E_{a,b}^k \longrightarrow E_{\pi_k(a),\pi_k(b)}^{k-1} \\ [X : Y : Z] \mapsto [\pi_k(X) : \pi_k(Y) : \pi_k(Z)] \end{array} \right. \tag{64}$$

is a surjective group homomorphism.
Proof.
π^k is well defined because π_k is a morphism of rings. According to theorem (2.7), we have $AY_{k-1} + BZ_{k-1} + CX_{k-1} = D \bmod p$, with

$$A = 2y_0 z_0, \tag{65}$$

$$B = y_0^2 - 3z_0^2 b_0 - 2z_0 a_0 x_0, \tag{66}$$

$$C = -\left(3x_0^2 + a_0 z_0^2\right) \tag{67}$$

and

$$D = b_{n-1} z_0^3 + a_{n-1} x_0 z_0^2. \tag{68}$$

The coefficients A, B and $-C$ are the partial derivatives of the function

$$F(X, Y, Z) = Y^2 Z - X^3 - a_0 X Z^2 - b_0 Z^3 \tag{69}$$

calculated starting from (x_0, y_0, z_0), which are not all equal to zero and deducing the existence of $\left[x_{k-1} : y_{k-1} : z_{k-1}\right]$. Hence, π^k est surjectif.
Using corollary (2.6), we deduce that π^k is a group homomorphism.
Lemma 4.2 *For all $k \geq 2$,*

$$Ker\left(\pi^k\right) = \left\{\left[l\epsilon^{k-1} : 1 : 0\right] \mid \ l \in \mathbb{F}_q\right\}. \tag{70}$$

Proof.
We have:

$$Ker\left(\pi^k\right) = \left\{P \in E_{a,b}^k \mid \ \pi^k(P) = [0 : 1 : 0] \ \right\}. \tag{71}$$

Then, $P = \left[x_{k-1}\epsilon^{k-1} : 1 + y_{k-1}\epsilon^{k-1} : z_{k-1}\epsilon^{k-1}\right] = \left[x_{k-1}\epsilon^{k-1} : 1 : z_{k-1}\epsilon^{k-1}\right]$..
As $P \in E_{a,b}^k$, we have

$$\begin{aligned} z_{k-1}\epsilon^{k-1} &= \left(x_{k-1}\epsilon^{k-1}\right)^3 + a x_{k-1}\epsilon^{k-1}\left(z_{k-1}\epsilon^{k-1}\right)^2 + b\left(z_{k-1}\epsilon^{k-1}\right)^3 \\ &= 0, \end{aligned} \tag{72}$$

so, $z_{k-1} = 0$.

This yields $Ker(\pi^k) = \{[l\epsilon^{k-1} : 1 : 0] | \quad l \in \mathbb{F}_q\}$.

Lemma 4.3 *The application*

$$\theta_k : \left| \begin{array}{lcl} \mathbb{F}_q & \rightarrow & E^k_{a,b} \\ l & \mapsto & [l\epsilon^{k-1} : 1 : 0] \end{array} \right. \tag{73}$$

is an injective group homomorphism.

Proof.

The application θ_k is injective by construction.

Let $[l\epsilon^{k-1} : 1 : 0]$ and $[h\epsilon^{k-1} : 1 : 0]$ be two elements in $E^k_{a,b}$, then:

$$\begin{aligned} k^\pi(l\epsilon^{k-1}) &= k^\pi(h\epsilon^{k-1}) \\ k^\pi(1) &= k^\pi(1) \\ k^\pi(0) &= k^\pi(0). \end{aligned} \tag{74}$$

so, using theorem (3.7),

$$\begin{aligned} X_3 &= (l+h)\epsilon^{k-1} \\ Y_3 &= 1 \\ Z_3 &= 0. \end{aligned} \tag{75}$$

This yields

$$\theta_k(l+h) = \theta_k(l) + \theta_k(h). \tag{76}$$

Thus, θ_k is an injective group homomorphism.

4.2 Main applications

In this subsection, we consider a prime p which does not divide N, where $N = \sharp E^1_{k^\pi(a),k^\pi(b)}$.

Corollary 4.4 *Let* $P \in E^k_{a,b}$*, then*

$$NP = [0 : 1 : 0] \Leftrightarrow P \in E^1_{k^\pi(a),k^\pi(b)}. \tag{77}$$

Proof.

If $P \in E^1_{k^\pi(a),k^\pi(b)}$, then $NP = [0 : 1 : 0]$.

Let $P = [x_0 + X : y_0 + Y : z_0 + Z] \in E^k_{a,b}$ and $Q = [x_0 : y_0 : z_0] \in E^1_{k^\pi(a),k^\pi(b)}$.

If $NP = [0 : 1 : 0]$, then $N(P - Q) = [0 : 1 : 0]$.

So, $P - Q = k_\theta(l_1, l_2, \ldots, l_{k-1})$.

We deduce that $Nl_i \equiv 0\ [p], i = 1, 2, \ldots, k-1$, where $pgcd(N, p) = 1$, which proves that $l_i = 0$ et $P = Q$.

Corollary 4.5

$$\forall P \in E^k_{a,b}, \textit{we have } pNP = [0 : 1 : 0]. \tag{78}$$

Proof.
$\forall P \in E^k_{a,b}, NP \in k_G,$ so $pNP = [0:1:0]$..
Lemma 4.6 *If p do not divide N, then there exists a unique homomorphism*

$$v_k : E^1_{a_0,b_0} \rightarrow E^k_{a,b} \tag{79}$$

for which the following diagram is commutative; (named Diagram(d)).

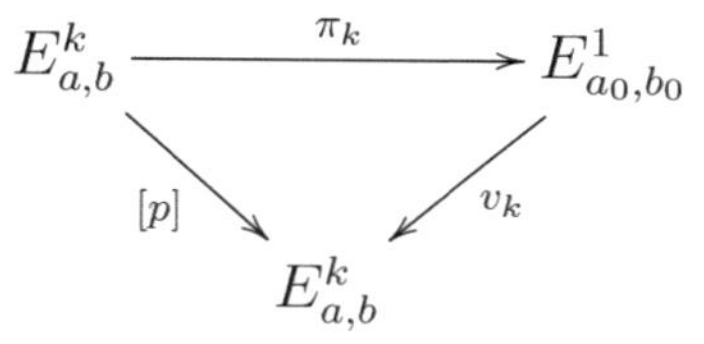

Proof.
Let $P \in k_G,$ we have $pP = [0:1:0]$.

Then

$$k_G \subset ker([p]). \tag{80}$$

Hence, there is a unique homomorphism

$$v_k : E^1_{a_0,b_0} \rightarrow E^k_{a,b} \tag{81}$$

which makes the diagram(d) commutative.
Theorem 4.7 *If p do not divide N, then there exists a unique homomorphism*

$$s_k : E^1_{a_0,b_0} \rightarrow E^k_{a,b} \tag{82}$$

such that $\pi_k o s_k = id_{E^1_{a_0,b_0}}$.
Proof.
Let $N' \in \mathbb{Z}$ as it exists $t \in \mathbb{Z}$ checking $1 - NN' = tp$. Then,

$$[1 - NN'] = [t]o[p]. \tag{83}$$

According to Lemma (4.6), there is a unique homomorphism

$$s_k : E^1_{a_0,b_0} \rightarrow E^k_{a,b} \tag{84}$$

which makes the following diagram commutative; (named Diagram(d')):

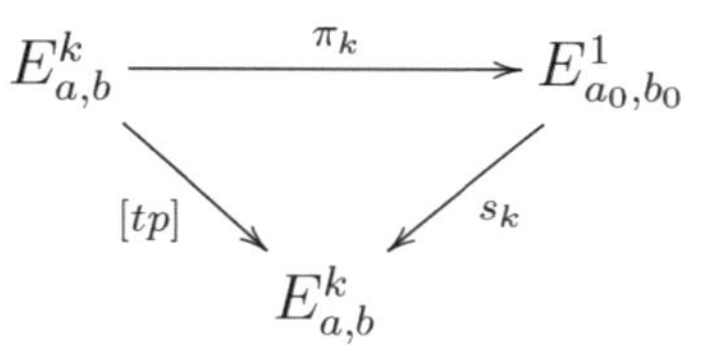

Let $P \in E^1_{a_0,b_0}$, then there exists $P' \in E^k_{a,b}$ such that $\pi_k(P') = P$. So,

$$\begin{aligned}\pi_k o s_k(P) &= \pi_k o s_k o \pi_k(P') \\ &= \pi_k([1 - NN'](P')) \\ &= \pi_k(P' - NN'P') \\ &= P - NN'P \\ &= P.\end{aligned} \tag{85}$$

Theorem 4.8 *If p do not divide N, then* $E^k_{a,b} \cong E^1_{a_0,b_0} \times k_G$.
Proof.
The isomorphism

$$f_k : \left| \begin{array}{ll} E^1_{a_0,b_0} \times k_G & \to E^k_{a,b} \\ (P, Q) & \mapsto s_k(P) + Q \end{array} \right. \tag{86}$$

admits an inverse application

$$F_k : \left| \begin{array}{l} E^k_{a,b} \to E^1_{a_0,b_0} \times k_G \\ P \mapsto (\pi_k(P), NN'P) \end{array} \right. . \tag{87}$$

Indeed,

$$\begin{aligned}f_k o F_k(P) &= f_k((\pi_k(P), NN'P)) \\ &= s_k o \pi_k(P) + NN'P \\ &= (1 - NN')P + NN'P \\ &= P.\end{aligned} \tag{88}$$

Likewise,

$$\begin{aligned}F_k o f_k(P, Q) &= F_k(s_k(P) + Q) \\ &= (\pi_k(s_k(P) + Q), NN'(s_k(P) + Q)).\end{aligned} \tag{89}$$

So,

$$\begin{aligned}\pi_k(s_k(P) + Q) &= \pi_k(s_k(P)) + \pi_k(Q) \\ &= P + [0 : 1 : 0] \\ &= P. \\ NN'(s_k(P) + Q) &= NN'(s_k(P)) + NN'Q \\ &= NN'(1 - NN')P' + NN'Q \\ &= N'tpNP' + NN'Q \\ &= [0 : 1 : 0] + NN'Q \\ &= NN'Q.\end{aligned} \tag{90}$$

As, $pQ = [0:1:0]$, then we have

$$\begin{aligned} NN'Q &= (1-tp)Q \\ &= Q - tpQ \\ &= Q. \end{aligned} \tag{91}$$

We conclude,

$$F_k o f_k(P,Q) = (P,Q). \tag{92}$$

Corollary 4.9 *If p do not divide N, then* $E^k_{a,b} \cong E^1_{a_0,b_0} \times \mathbb{F}_q^{k-1}$.
Proof.
We have, $k_G \cong \mathbb{F}_q^{k-1}$, see [27, 30, 33].
Corollary 4.10 *If p do not divide N, then*

$$\begin{aligned} &E^k_{a,b} \cong C_N \times \mathbb{F}_q^{k-1}, \textit{with } C_N \textit{ cyclic} \\ &\quad or \\ &E^k_{a,b} \cong \mathbb{Z}/n_1\mathbb{Z} \times \mathbb{Z}/n_2\mathbb{Z} \times \mathbb{F}_q^{k-1}, \textit{where } n_2|(n_1 \wedge p - 1). \end{aligned} \tag{93}$$

Proof.
We have

$$\begin{aligned} &E^1_{a_0,b_0} \cong C_N, \text{with } C_N \text{ cyclic} \\ &\quad or \\ &E^1_{a_0,b_0} \cong \mathbb{Z}/n_1\mathbb{Z} \times \mathbb{Z}/n_2\mathbb{Z}, \text{where } n_2|(n_1 \wedge p - 1). \end{aligned} \tag{94}$$

And

$$E^k_{a,b} \cong E^1_{a_0,b_0} \times \mathbb{F}_q^{k-1}. \tag{95}$$

Corollary 4.11 *If p do not divide N, then* $(\sqrt{q}-1)^2 q^{k-1} \leq \sharp(E^k_{a,b}) \leq (\sqrt{q}+1)^2 q^{k-1}$.
Proof.
According to Haas' theorem, we have:

$$|q+1-N| \leq 2\sqrt{q} \tag{96}$$

so

$$(\sqrt{q}-1)^2 q^{k-1} \leq \sharp(E^k_{a,b}) \leq (\sqrt{q}+1)^2 q^{k-1}. \tag{97}$$

5. Applications

In this section, we are interested in ECC using elliptic curves over the ring R_n.

5.1 The discrete logarithm on $E^n_{a,b}$

The discrete logarithm problem that we denote DLP, (Discrete logarithm problem), is a generally difficult problem which depends on the considered group $\mathcal{G}$. In many situations, due to the asymmetry existing between problems concerning the calculation of logarithms and calculation of powers which is more easier and so of great interest in cryptograph, the above mentioned makes Diffie and Hellman were the first to build a cryptosystem from this situation [37, 38].

Definition 5.1 *Let $\mathcal{G}$ be a finite cyclic group of order ρ and s, r two elements of $\mathcal{G}$. We call discrete logarithm of base s of r, the only element m in $[[0, \rho - 1]]$ such that $s^m = r$. The discrete logarithm for elliptic curves is defined in an analogous way to be, the only element m in $[[0, \rho - 1]]$ such that $mP = Q$, where P, and Q are two points of an additive subgroup $\mathcal{G}$ of $\mathbf{E}^n_{a,b}$.*

By using the isomorphism proved given in theorem (4.8), we get the results gathered in the next theorem:

Theorem 5.2 *If p does not divide N, then.*

- $\#\mathbf{E}^n_{a,b} = p^{d(n-1)} \times N$.
- The problem of the discrete logarithm on the elliptic curve $\mathbf{E}^n_{a,b}$ is equivalent to that of $\mathbf{E}^1_{a_0,b_0}$.
- If the problem of the discrete logarithm on $\mathbf{E}^n_{a,b}$ is trivial, then it is also trivial on the elliptic curve $\mathbf{E}^1_{a_0,b_0}$.

5.2 Cryptography based on elliptic curves $\mathbf{E}^n_{a,b}$

Elliptic curve cryptography (ECC) is public key cryptography, which relies on the use of curves over finite fields. Essentially, there are two families of these curves which are used in cryptography. The first uses elliptic curves on a finite field $\mathbb{F}_{p^d}$, where p is a large prime number. This family is the best choice for a high software level when implementing ECC. The second family uses elliptic curves on a binary field $\mathbb{F}_{2^d}$ where d is a large positive integer, this family is more appropriate at the material level point of view when implementing ECC. Another family which is also interesting in ECC implementations is the family of elliptic curves on the previously seen rings $\mathbf{R_n}$. The most important advantage presented by the use of elliptic curves in cryptography (ECC) consists in the high security they provide for wireless applications compared to other asymmetric key cryptosystems, also their small key size. Indeed, a 160-bit key for (ECC) can replace a 1024-bit key for (RSA). Given d; a large integer, $P \in \mathbf{E}^n_{a,b}$ and $Q \in \mathcal{G} \subset \mathbf{E}^n_{a,b}$. The discrete elliptical logarithm problem (DLEP) consists in finding $k \in \mathbb{Z}\, such\ \ that\ Q = [k]P$, where

$$[k]P = \underbrace{P + P + \cdots P}_{ktimes} = kP. \tag{98}$$

This is in fact a difficult problem, whose resolution is exponential.

5.3 Elliptical Diffie-Hellman cryptosystem

Recall that Alice and Bob can publicly agree on a common secret (that we describe below).

1. They choose on a large integer d, $\mathbf{E}^n_{a,b}$ *and* $P \in \mathbf{E}^n_{a,b}$.

2. Alice chooses $t \in \mathbb{Z}$ and calculates tP.

3. Bob chooses $s \in \mathbb{Z}$ and calculates sP.

4. Alice let public tP and keep private t.

5. Bob let public sP and keep private s.

6. Then, Alice and Bob build their common secret key $K = tsP = stP$.

Remark 5.3

1. Unlike the classic Diffie-Hellman algorithm, we do not ask that P be a generator of $\mathbf{E}^n_{a,b}$. The analogue of the subgroup $\mathbb{F}^*_p$ of order $p-1$, is the cyclic subgroup of $\mathbf{E}^n_{a,b}$, generated by the point P.

2. As soon as we have a group $\mathbf{E}^n_{a,b}$, and an element $P \in \mathbf{E}^n_{a,b}$ of finite order we can consider a Diffie-Hellman system on $\mathcal{G} = <P>$ which is cyclic. For this construction to have a cryptographic interest, $log_P(tP) = t$ must be not easy to calculate.

3. $\mathbf{E}^n_{a,b}$, is not always cyclical.

4. If, Oscar (program) is giving $d, \mathbf{E}^n_{a,b}, tP\ and\ sP$, then it is able to solve the discrete elliptical logarithm problem and find $t\ or\ s$.

5.4 Elliptical ElGamal cryptosystem

Let $P_m \in \mathbf{E}^n_{a,b}$ be the point representing the message m, to encrypt P_m :

1. Key generation algorithm

    - Bob chooses the private key $t \in \mathbb{Z}$ known only to him.

    - $d \in \mathbb{N}, P \in \mathbf{E}^n_{a,b}\ and\ R = [t]P$ are public.

2. Encryption algorithm

    - Alice Randomly chooses $k \in \mathbb{Z}$;

    - She calculates $c_1 = [k]P \in \mathbf{E}^n_{a,b}$;

    - She also calculates $c_2 = P_m + [k]R$;

    - Then, he makes public c_1, c_2, or $C = (c_1; c_2)$.

3. To decrypt received message (c_1, c_2), Bob calculates:

$$P_m = c_2 - [k]R = c_2 - [k][t]P = c_2 - [t]c_1. \tag{99}$$

Now, Oscar encounters the discrete elliptic logarithm problem, because to decipher the message P_m he must know t (i.e.; calculate t such that $R = [t]P$).

5.5 Coding example

Let d be a positive integer, we consider the quotient ring $\mathbf{R}_2 = \frac{\mathbb{F}_{2^d}[X]}{(X^2)}$, where $\mathbb{F}_{2^d}$ is the finite field of order 2^d.

Then the ring $\mathbf{R}_2$ is identified with the ring $\mathbb{F}_{2^d}[\varepsilon]$, where $\varepsilon^2 = 0$, i.e.,

$$\mathbf{R}_2 = \{a_0 + a_1 \cdot \varepsilon | a_0, a_1 \in \mathbb{F}_{2^d}\}. \quad (100)$$

We consider the elliptic curve on the ring $\mathbf{R}_2$ given by the equation:

$$Y^2Z + XYZ = X^3 + aX^2Z + bZ^3. \quad (101)$$

where a, b in $\mathbf{R}_2$ and b is invertible in $\mathbf{R}_2$. Each element of $\mathbf{E}^2_{a,b}$ is of the form; $[X : Y : 1]$ or $[x\varepsilon : 1 : 0]$, with $x \in \mathbb{F}_{2^d}$. Write:

$$\mathbf{E}^2_{a,b} = \{[X : Y : 1] \in \mathbb{P}^2_2 | Y^2 + XY = X^3 + aX^2 + b\} \cup \{[x\varepsilon : 1 : 0] | x \in \mathbb{F}_{2^d}\}. \quad (102)$$

Let $\mathbf{E}^2_{a,b}$ be the elliptic curve over R_2 and consider the irreducible polynomial $T(X) = 1 + X + X^3$ in $\mathbb{F}_2[X]$. Let α be such that $T(\alpha) = 0$ in $\mathbb{F}_8 = \frac{\mathbb{F}_2[X]}{(T(X))}$, then $(1, \alpha, \alpha^2)$ is a vector space base of $\mathbb{F}_8$ over $\mathbb{F}_2$.

$$\mathbb{F}_8 = \{0, 1, \alpha, \alpha^2, \alpha + 1, \alpha^2 + \alpha, \alpha^2 + 1, \alpha^2 + \alpha + 1\} \quad (103)$$

⋆ Put:

$$a = 1 + \alpha; \quad (104)$$

$$b = 1 + \alpha^2\varepsilon. \quad (105)$$

We have: $\mathbf{R}_2 = \mathbb{F}_8[\varepsilon]$ and $\mathbf{E}^2_{a,b} : Y^2 + XY = X^3 + (1 + \alpha)X^2 + (1 + \alpha^2\varepsilon)$. Consider $P \in \mathbf{E}^2_{a,b}$ of order l, and consider the subgroup $\mathcal{G} = <P>$, generated by P, to encrypt and decrypt our messages.

1. Coding of elements of $\mathcal{G}$

We will give a code to each element $Q = mP$, where $m \in \{1, 2, .., l\}$, defined as follows:
If $Q = [x_0 + x_1\varepsilon : y_0 + y_1\varepsilon : Z]$, where $x_i, y_i \in \mathbb{F}_8$ for $i = 0, 1$, and $Z = 0\, or\, 1$, then we set:

$$x_i = c_{0i} + c_{1i}\alpha + c_{2i}\alpha^2; \quad (106)$$

$$y_i = d_{0i} + d_{1i}\alpha + d_{2i}\alpha^2, \quad (107)$$

where α is the primitive root of the irreducible polynomial $T(X) = 1 + X + X^3$, and $c_{ij}, d_{ij} \in \mathbb{F}_2$.
So, we code Q as follows:
If $Z = 1$: $Q = c_{00}c_{10}c_{20}c_{01}c_{11}c_{21}d_{00}d_{10}d_{20}d_{01}d_{11}d_{21}1$.
If $Z = 0$: $Q = 00c_{01}c_{11}c_{21}d_{01}d_{11}d_{21}10000$.

2. Example: with the same a and b;

$$a = 1 + \alpha; \tag{108}$$

$$b = 1 + \alpha^2\varepsilon. \tag{109}$$

The elliptic curve $\mathbf{E}^2_{a,b}$ contains 112 elements to know:

$$\mathbf{E}_{a,b}(\mathbb{A}_2)=\{[(\alpha^2+\alpha+1)\varepsilon:1:0],[(\alpha^2+1)\varepsilon:1:0],[(1+\alpha)\varepsilon:1:0],[(\alpha^2+\alpha)\varepsilon:1:0],[\alpha^2\varepsilon:1:0],[\alpha\varepsilon:1:0],[\varepsilon+\alpha^2+1:\alpha\varepsilon+\alpha:1],[\alpha^2+\alpha:(\alpha^2+\alpha+1)\varepsilon:1],[\alpha^2\varepsilon:\varepsilon+1:1],[(\alpha^2+1)\varepsilon+\alpha^2+1:(\alpha^2+\alpha)\varepsilon+\alpha:1],[(1+\alpha)\varepsilon+\alpha^2:(\alpha^2+\alpha+1)\varepsilon+\alpha^2+1:1],[(\alpha^2+\alpha)\varepsilon+1+\alpha:(\alpha^2+\alpha+1)\varepsilon+\alpha^2:1],[(1+\alpha)\varepsilon+\alpha^2+1:\alpha:1],[\alpha^2\varepsilon+\alpha^2+\alpha+1:\varepsilon+1:1],[1+\alpha:(\alpha^2+1)\varepsilon+\alpha^2+\alpha+1:1],[(\alpha^2+\alpha+1)\varepsilon+\alpha:(\alpha^2+1)\varepsilon+\alpha^2+\alpha:1],[\alpha\varepsilon+\alpha^2:(\alpha^2+1)\varepsilon+\alpha^2+1:1],[(1+\alpha)\varepsilon+\alpha^2+\alpha+1:(1+\alpha)\varepsilon+1:1],[\alpha^2\varepsilon:(1+\alpha)\varepsilon+1:1],[\alpha^2\varepsilon+1+\alpha:\alpha^2\varepsilon+\alpha^2+\alpha+1:1],[\alpha\varepsilon+\alpha^2+1:\varepsilon+\alpha:1],[(1+\alpha)\varepsilon+\alpha^2+\alpha:(\alpha^2+1)\varepsilon+\alpha^2+\alpha:1],[\alpha^2\varepsilon+\alpha^2+\alpha+1:(\alpha^2+1)\varepsilon+\alpha^2+\alpha:1],[(\alpha^2+1)\varepsilon+1+\alpha:(\alpha^2+\alpha)\varepsilon+\alpha^2:1],[\alpha\varepsilon+\alpha^2+\alpha: \tag{110}$$

$$0:1],[\alpha^2+1:(1+\alpha)\varepsilon+\alpha^2+\alpha+1:1],[\varepsilon+\alpha^2+1:(1+\alpha)\varepsilon+\alpha^2+\alpha+1:1],[(\alpha^2+\alpha+1)\varepsilon+\alpha^2+1:(1+\alpha)\varepsilon+\alpha^2+\alpha+1:1],[(\alpha^2+\alpha+1)\varepsilon+\alpha^2:(1+\alpha)\varepsilon+1:1],[\alpha^2\varepsilon+\alpha^2+1:(\alpha^2+\alpha+1)\varepsilon+\alpha:1],[(1+\alpha)\varepsilon+\alpha^2+\alpha:(\alpha^2+\alpha)\varepsilon:1],[(\alpha^2+\alpha+1)\varepsilon+1+\alpha:\varepsilon+\alpha^2:1],[(\alpha^2+\alpha)\varepsilon+\alpha^2+1:(1+\alpha)\varepsilon+\alpha^2+\alpha+1:1],[(\alpha^2+1)\varepsilon+\alpha^2+1:(1+\alpha)\varepsilon+\alpha^2+\alpha+1:1],[(\alpha^2+1)\varepsilon+1+\alpha:(1+\alpha)\varepsilon+\alpha^2+\alpha+1:1],[\varepsilon+\alpha^2:\alpha\varepsilon+1:1],[(\alpha^2+1)\varepsilon+\alpha^2+\alpha+1:\alpha\varepsilon+1:1],[\alpha^2\varepsilon:\alpha\varepsilon+1:1],[\varepsilon+\alpha^2+\alpha:\varepsilon:1],[(1+\alpha)\varepsilon+\alpha^2+1:(1+\alpha)\varepsilon+\alpha^2+\alpha+1:1],[\alpha^2\varepsilon+\alpha^2+1:(1+\alpha)\varepsilon+\alpha^2+\alpha+1:1],[\alpha\varepsilon+\alpha^2+1:(1+\alpha)\varepsilon+\alpha^2+\alpha+1:1],[(\alpha^2+\alpha+1)\varepsilon+\alpha^2+\alpha+1:\alpha^2\varepsilon+1:1],[(1+\alpha)\varepsilon+\alpha^2:\alpha^2\varepsilon+1:1],[\alpha^2\varepsilon:\alpha^2\varepsilon+1:1],[\alpha^2+\alpha+1:(\alpha^2+\alpha)\varepsilon+\alpha^2+\alpha:1],[\alpha^2\varepsilon:1:1],[(\alpha^2+\alpha)\varepsilon+\alpha^2:1:1],[\varepsilon+\alpha,(1+\alpha)\varepsilon+\alpha^2+\alpha:1],[(\alpha^2+\alpha+1)\varepsilon+\alpha^2+\alpha+1:(1+\alpha)\varepsilon+\alpha^2+\alpha:1],[(\alpha^2+\alpha)\varepsilon+\alpha^2+\alpha:(1+\alpha)\varepsilon+\alpha^2+\alpha:1],[(\alpha^2+\alpha)\varepsilon+1+\alpha:\varepsilon+\alpha^2+\alpha+1:1],[(\alpha^2+\alpha+1)\varepsilon+\alpha^2+\alpha:(1+\alpha)\varepsilon:1],[\varepsilon+\alpha^2+\alpha+1:\alpha^2\varepsilon+\alpha^2+\alpha:1],[\varepsilon+1+\alpha:\alpha\varepsilon+\alpha^2+\alpha+1:1],[\alpha:\alpha\varepsilon+\alpha^2+\alpha:1],[\alpha\varepsilon+\alpha^2+\alpha+1:\alpha\varepsilon+\alpha^2+\alpha:1],[\alpha\varepsilon+\alpha^2+\alpha+1:1:1],[(\alpha^2+\alpha+1)\varepsilon+\alpha^2+\alpha:\alpha^2\varepsilon+\alpha^2+\alpha:1],[\alpha^2:\varepsilon+\alpha^2+1:1],[(1+\alpha)\varepsilon+1+\alpha:(\alpha^2+\alpha+1)\varepsilon+\alpha^2+\alpha+1:1],[\alpha^2+\alpha:(\alpha^2+\alpha+1)\varepsilon+\alpha^2+\alpha:1],[\varepsilon+\alpha^2+\alpha:\alpha^2+\alpha:1],[\alpha\varepsilon+1+\alpha:\alpha^2+\alpha+1:1],[\alpha^2:\varepsilon+1:1],[(\alpha^2+\alpha)\varepsilon+\alpha^2+\alpha+1:(\alpha^2+\alpha+1)\varepsilon+1:1],[\alpha^2\varepsilon:(\alpha^2+\alpha+1)\varepsilon+1:1],[(\alpha^2+\alpha)\varepsilon+\alpha^2+1:(\alpha^2+1)\varepsilon+\alpha:1],[\varepsilon+\alpha^2:(1+\alpha)\varepsilon+\alpha^2+1:1],[\varepsilon+1+\alpha:(1+\alpha)\varepsilon+\alpha^2:1],[(\alpha^2+\alpha)\varepsilon+\alpha^2:(\alpha^2+\alpha)\varepsilon+\alpha^2+1:1],[(\alpha^2+\alpha)\varepsilon+\alpha^2+\alpha:(\alpha^2+1)\varepsilon:1],[\varepsilon+\alpha^2+\alpha+1:(\alpha^2+1)\varepsilon+1:1],[(\alpha^2+\alpha+1)\varepsilon+\alpha^2+1:\alpha^2\varepsilon+\alpha:1],[\alpha^2\varepsilon+\alpha^2+\alpha:\alpha\varepsilon:1],[(\alpha^2+\alpha)\varepsilon+\alpha:\alpha^2\varepsilon+\alpha^2+\alpha:1],[(\alpha^2+\alpha)\varepsilon+\alpha:\alpha\varepsilon+\alpha^2:1],[(\alpha^2+1)\varepsilon+\alpha:\alpha\varepsilon+\alpha^2:1],[(1+\alpha)\varepsilon+\alpha:\alpha\varepsilon+\alpha^2:1],[(1+\alpha)\varepsilon+\alpha:\varepsilon+\alpha^2+\alpha:1],[(1+\alpha)\varepsilon+\alpha^2+\alpha+1:\alpha^2+\alpha:1],[\alpha^2+\alpha+1:(\alpha^2+\alpha)\varepsilon+1:1],[\alpha^2\varepsilon:(\alpha^2+\alpha)\varepsilon+1:1],[\alpha^2\varepsilon+\alpha:\alpha\varepsilon+\alpha^2:1],[\alpha\varepsilon+1+\alpha:\alpha\varepsilon+\alpha^2:1],[\alpha\varepsilon+\alpha:\alpha\varepsilon+\alpha^2:1],[(\alpha^2+1)\varepsilon+\alpha^2:\alpha^2+1:1],[\alpha\varepsilon+\alpha:\alpha^2+\alpha:1],[(1+\alpha)\varepsilon+1+\alpha:\alpha^2\varepsilon+\alpha^2:1],[(\alpha^2+1)\varepsilon+\alpha^2+\alpha:\varepsilon+\alpha^2+\alpha:1],[\alpha\varepsilon+\alpha^2+\alpha:\alpha\varepsilon+\alpha^2+\alpha:1],[\alpha^2\varepsilon+\alpha^2:\alpha\varepsilon+\alpha^2+1:1],[(\alpha^2+1)\varepsilon+\alpha^2+\alpha+1:(\alpha^2+\alpha+1)\varepsilon+\alpha^2+\alpha:1],[(\alpha^2+1)\varepsilon+\alpha:(\alpha^2+\alpha+1)\varepsilon+\alpha^2+\alpha:1],[(\alpha^2+\alpha+1)\varepsilon+\alpha^2:\alpha^2\varepsilon+\alpha^2+1:1],[\alpha^2\varepsilon: \tag{111}$$

$$(\alpha^2+1)\varepsilon+1:1], [\alpha^2\varepsilon+\alpha:(\alpha^2+\alpha)\varepsilon+\alpha^2+\alpha:1], [1+\alpha:(\alpha^2+1)\varepsilon+\alpha^2: 1], [(\alpha^2+\alpha+1)\varepsilon+1+\alpha:(\alpha^2+\alpha)\varepsilon+\alpha^2+\alpha+1:1], [\alpha^2\varepsilon+\alpha^2:(\alpha^2+\alpha)\varepsilon+1: 1], [\alpha^2+1:(1+\alpha)\varepsilon+\alpha:1], [(\alpha^2+1)\varepsilon+\alpha^2+\alpha:\alpha^2\varepsilon:1], [\alpha^2\varepsilon+1+\alpha:\alpha^2: 1], [0:1:0], [\alpha:\alpha\varepsilon+\alpha^2:1], [\varepsilon+\alpha:\alpha\varepsilon+\alpha^2:1], [(\alpha^2+\alpha+1)\varepsilon+\alpha:\alpha\varepsilon+\alpha^2: 1], [(\alpha^2+\alpha)\varepsilon+\alpha^2+\alpha+1:\varepsilon+\alpha^2+\alpha:1], [\alpha^2\varepsilon+\alpha^2+\alpha:(\alpha^2+\alpha)\varepsilon+\alpha^2+\alpha: 1], [\alpha\varepsilon+\alpha^2:(\alpha^2+\alpha+1)\varepsilon+1:1], [\varepsilon:1:0], [(\alpha^2+1)\varepsilon+\alpha^2:(\alpha^2+1)\varepsilon+1:1]\} \tag{112}$$

We consider: $P=[\alpha:\alpha+\alpha^2+\alpha\varepsilon:1]=0100000110101$, then $\mathcal{G}=<P>$ is of order 28. We attach to each point $Q\in\mathcal{G}$ a letter of the alphabet and a code. We collect the results in the following **Table 2**:

	mP	Codefor mP	Symbol
1	$[\alpha:\alpha+\alpha^2+\alpha\varepsilon:1]$	0100000110101	*a*
2	$[1+\alpha+\varepsilon:\alpha^2+(1+\alpha)\varepsilon:1]$	1101000011101	*b*
3	$[\alpha^2+\alpha\varepsilon:1+\alpha^2+(\alpha^2+1)\varepsilon:1]$	0010101011011	*c*
4	$[1+\alpha+\alpha^2+(1+\alpha)\varepsilon:\alpha+\alpha^2:1]$	1111100110001	*d*
5	$[\alpha+\alpha^2+(1+\alpha+\alpha^2)\varepsilon:(1+\alpha)\varepsilon:1]$	0111110001101	*e*
6	$[1+\alpha^2+(1+\alpha+\alpha^2)\varepsilon:\alpha+\alpha^2\varepsilon:1]$	1011110100011	*f*
7	$[\alpha^2\varepsilon:1+(1+\alpha^2)\varepsilon:1]$	0000011001011	*g*
8	$[1+\alpha^2+(1+\alpha^2)\varepsilon:1+\alpha+\alpha^2+(1+\alpha)\varepsilon:1]$	1011011111101	*h*
9	$[\alpha+\alpha^2+\alpha\varepsilon:\alpha+\alpha^2+\alpha\varepsilon:1]$	0110100110101	*i*
10	$[1+\alpha+\alpha^2+\alpha\varepsilon:1:1]$	1110101000001	*j*
11	$[\alpha^2+\alpha^2\varepsilon:1+(\alpha+\alpha^2)\varepsilon:1]$	0010011000111	*k*
12	$[1+\alpha+(\alpha+\alpha^2)\varepsilon:1+\alpha+\alpha^2+\varepsilon:1]$	1100111111001	*l*
13	$[\alpha+(1+\alpha)\varepsilon:\alpha^2+\alpha\varepsilon:1]$	0101100010101	*m*
14	$[\alpha^2\varepsilon:1:0]$	0000100010000	*n*
15	$[\alpha+(1+\alpha)\varepsilon:\alpha+\alpha^2+\varepsilon:1]$	0101100111001	*o*
16	$[1+\alpha+(\alpha+\alpha^2)\varepsilon:\alpha^2+(1+\alpha+\alpha^2)\varepsilon:1]$	1100110011111	*p*
17	$[\alpha^2+\alpha^2\varepsilon:1+\alpha^2+\alpha\varepsilon:1]$	0010011010101	*q*
18	$[1+\alpha+\alpha^2+\alpha\varepsilon:\alpha+\alpha^2+\alpha\varepsilon:1]$	1110100110101	*r*
19	$[\alpha+\alpha^2+\alpha\varepsilon:0:1]$	0110100000001	*s*
20	$[1+\alpha^2+(1+\alpha^2)\varepsilon:\alpha+(\alpha+\alpha^2)\varepsilon:1]$	1011010100111	*t*
21	$[\alpha^2\varepsilon:1+\varepsilon:1]$	0000011001001	*u*
22	$[1+\alpha^2+(1+\alpha+\alpha^2)\varepsilon:1+\alpha+\alpha^2+(1+\alpha)\varepsilon:1]$	1011111111101	*v*
23	$[\alpha+\alpha^2+(1+\alpha+\alpha^2)\varepsilon:\alpha+\alpha^2+\alpha^2\varepsilon:1]$	0111110110011	*w*
24	$[1+\alpha\alpha^2+(1+\alpha)\varepsilon:1+(1+\alpha)\varepsilon:1]$	1111101001101	*x*
25	$[\alpha^2+\alpha\varepsilon:1+(1+\alpha+\alpha^2)\varepsilon:1]$	0010101001111	*y*
26	$[1+\alpha+\varepsilon:1+\alpha+\alpha^2+\alpha\varepsilon:1]$	1101001110101	*z*
27	$[\alpha:\alpha^2+\alpha\varepsilon:1]$	0100000010101	*space*
28	$[0:1:0]$	0000000010000	,

Table 2.
Points coding.

5.6 Encryption and decryption procedures

- The encryption of our message "for the elliptical curve", is;

$$\begin{array}{l} 01111101100110111110001101010000001010100000 \\ 110010010110100000001011111000110101000000010 \\ 1011011010100111101101111101011111000110101 \\ 00000010101011111000110111001111110011100111 \\ 1110010110100110101110011001111101101010011 \\ 10110100110101001010101101101000000101010010 \\ 1010110110000011001001111010011010110111111 1 \\ 110101111100011010000000010000 \end{array} \tag{113}$$

- Decryption of this message:

$$\begin{array}{l} 010000011010111010000111011111100110001011 11 \\ 1000110111001111110011011011111101010000011 0 \\ 10101011000101010110100110101111110011000101 \\ 000000101011011010100111010000011010111111 00 \\ 110001010110001010101011001110011110100110101 0 \\ 110100110101 \end{array} \tag{114}$$

is: hello to abdelhakim.

6. Conclusion

The results obtained are very important from theoretical points of view because to study an elliptic curve on a finite local ring it suffices to study these curves on finite fields, for the applications of these curves they can be applied in cryptography to reinforce security and we can use them in cryptanalysis to solve the PDL on special curves. This results are very imploring and give applications in different fields such as classical mechanics, number theory, cryptology, information theory … and we can quote here:

1. The generalization of Hass's theorem, corollary 4.9.

2. The result of the corollary 4.11, then in [24], we have the result of the Proposition 3.12.

Acknowledgements

The authors thank professor Karim Mounirh for his correcting typographical, grammatical, spelling, and punctuation errors to ensure the clear and accurate communication.

Author details

Abdelhakim Chillali* and Lhoussain El Fadil
Sidi Mohamed Ben Abdellah University, Morocco

*Address all correspondence to: abdelhakim.chillali@usmba.ac.ma

References

[1] Boneh D, Franklin M. Identity-based encryption from the Weil pairing. SIAM Journal on Computing. 2003;**32**(3): 586-615

[2] Hoffstein J, Pipher J, Silverman JH. An Introduction to Mathematical Cryptography. Springer, New York, NJ, USA: Undergraduate Texts in Mathematics; 2008

[3] Koblitz N. Elliptic curve cryptosystems. Mathematics of Computation. 1987;**48**:203-209

[4] Diffie W, Hellman ME. New directions in cryptography. In: IEEE Transactions on Information Theory. Vol. 22. IEEE Xplore; 1976. pp. 644-654

[5] Johnson D, Menezes A, Vanstone S. The elliptic curve digital signature algorithm (ECDSA). International Journal of Information Security. 2001;**1**(1):36-63

[6] Tzer-Shyong C, Gwo-Shiuan H, Tzuoh-Pyng L, Yu-Fang C. Digital signature scheme resulted from identification protocol by elliptic curve cryptosystem. In: Tencon '02. Proceedings. 2002 IEEE Region 10 Conference on Computers, Communications, Control and Power Engineering. Vol. 1. Beijing: IEEE; 2002. pp. 192-195

[7] Shannon CE. Communication theory of secrecy systems. Bell System Technical Journal. 1949;**28**(4):656-715

[8] Shannon CE. A mathematical theory of communication. Bell System Technical Journal. 1948;**27**(4):379-423

[9] Beutelspacher A, Rosenbaum U. Projective Geometry: From Fondations to Application. Cambridge: Cambridge University Press; 1998

[10] Hartshorne J. Algebraic Geometry. Vol. 52. Springer-Verlag, GTM; 1977

[11] Lang S. Algebraic Number Theory. New York: Springer; 1986

[12] Lenstra AK, Lenstra HW Jr, editors. The Development of the Number Field Sieve, Volume 1554 of Lecture Notes in Mathematics. Berlin: Springer-Verlag; 1993

[13] Joux A. A one round protocol for tripartite Diffie-Hellman. Journal of Cryptology. 2004;**17**(4):263-276

[14] Joux A, Pierrot C. Improving the Polynomial Time Precomputation of Frobenius Representation Discrete Logarithm Algorithms Simplified Setting for Small Characteristic Finite Fields. Springer-Verlag; 2014

[15] Joux A, Vitse V. Elliptic curve discrete logarithm problem over small degree extension fields application to the static Diffie-Hellman problem on $\mathbf{E}(\mathbb{F}_{q^5})$. Journal of Cryptology. 2013;**26**(1):119-143

[16] Maurer UM, Wolf S. Diffie Hellman oracles. In: Crypto, 96, LNCS. 1109. Springer-Verlag; 1996. pp. 268-282

[17] Menezes AJ, Okamoto T, Vanstone SA. Reducing elliptic curve logarithms to logarithms in a finite field. IEEE Transactions on Information Theory. 1993;**39**(5):1639-1646

[18] Popescu C. An identification scheme based on the elliptic curve discrete logarithm problem. In: Proceedings of the Fourth International Conference/ Exhibition on High Performance Computing in the Asia-Pacific Region. Vol. 2. Beijing: IEEE; 2000. pp. 624-625

[19] Fontein F. Elliptic Curves over Rings with a Point of View on Cryptography and Factoring. Vol. 28. Company, Reading, Massachusetts:

Diplomarbeit, Carl von Ossietzky Universität Oldenburg Diplomstudiengang Mathematik; 2005

[20] Silverman JH. Advanced topics in the arithmetic of elliptic curves. In: Volume 151 of Graduate Texts in Mathematics. Springer; 1994

[21] Silverman JH. The arithmetic of elliptic curves. In: Graduate Texts in Mathematics. Vol. 106. Springer; 1986

[22] Semaev I. New algorithm for the discrete logarithm problem on elliptic curves. Computer Science and Security. 2015; Cryptology ePrint Archive, Report 2015/310

[23] Elhassani M, Chillali A, Mouhib A. Elliptic curve and lattice cryptosystem. In: Proceedings - 2019 International Conference on Intelligent Systems and Advanced Computing Sciences. ISACS; 2019

[24] Boulbot A, Chillali A, Mouhib A. Elliptic curves over the ring R. Boletim da Sociedade Paranaense de Matematica (BSPM). 2020;**38**(3):193-201

[25] Sahmoudi M, Chillali A. Elliptic curve on a family of finite ring. WSEAS Transactions on Mathematics. 2019;**18**: 415-422

[26] Sahmoudi M, Chillali A. SCC-cryptosystem on an algebraic closure ring. Journal of Discrete Mathematical Sciences and Cryptography. 2019. DOI: 10.1080/09720529.2019.1624338

[27] Chillali A. Identification methods over. In: ICACM'11 Proceedings of the 2011 International Conference on Applied & Computational Mathematics, ACM Digital Library. Recent Researches in Applied and Computational Mathematics. 2011. pp. 133-138

[28] Hassib MH, Chillali A, Elomary MA. Special ideal ring A3 and cryptography. In: Proceedings of the International Conference JNS3. IEEE; 2013. pp. 1-4

[29] Tadmori A, Chillali A, Ziane M. The binary operations calculus in $\mathbf{E}_{a,b,c}$. International Journal of Mathematical Models and Methods in Applied Sciences. 2015;**9**:171-175

[30] Chillali A. The $J_{\varepsilon,n}$-invariant of $\mathbf{E}^n_{A,B}$. In: Recent Advances in Computers, Communications, Applied Social Science and Mathematics -Proceedings of ICANCM'11, ICDCC'11, IC-ASSSE-DC'11. 2011

[31] Tadmori A, Chillali A, Ziane M. Elliptic curve over ring $\mathbf{A_4} = \mathbb{F}_{2^d}[\varepsilon]; \varepsilon^4 = 0$. Applied Mathematical Sciences. 2015;**9**(35):1721-1733

[32] Bosma W, Lenstra H. Complete system of two addition laws for elliptic curved. Journal of Number Theory. 1995;**53**:229-240

[33] Chillali A. Cryptography over elliptic curve of the ring $\mathbb{F}_q[\varepsilon], \varepsilon^4 = 0$. World Academy of Science, Engineering and Technology. 2011;**5**:917-919

[34] Hassib MH, Chillali A, Elomary MA. Elliptic curves over a chain ring of characteristic 3. Journal of Taibah University for Science. 2014;**9**(3): 276-287

[35] Tadmori A, Chillali A, Ziane M. Normal form of the elliptic curve over the finite ring. Journal of Mathematics and System Science. 2014;**4**:194-196

[36] Tadmori A, Chillali A, Ziane M. Elliptic curve over SPIR of characteristic two. In: Proceeding of the 2013 International Conference on Applied Mathematics and Computational Methodes. 2013. pp. 41-44

[37] Okamoto T, Uchiyama S. A new pubic-key cryptosystem as secure as factoring. In: Eurocrypt, 98, LNCS. 1403. Springer-Verlag; 1998. pp. 308-318

[38] Washington LC. Elliptic curves number theory and cryptography. In: Discrete Mathematics and its Applications. Chapman and Hall/CRC; 2003